P9-DMA-138

SEEING WHAT'S NEXT

SEEING WHAT'S NEXT

USING THE THEORIES OF INNOVATION TO PREDICT INDUSTRY CHANGE

CLAYTON M. CHRISTENSEN
SCOTT D. ANTHONY
ERIK A. ROTH

HARVARD BUSINESS SCHOOL PRESS
BOSTON, MASSACHUSETTS

Printed in the United States of America
08 07 06 05 04   5 4 3 2 1

Library of Congress Cataloging-in-Publication Data
Christensen, Clayton M.
Seeing what's next? : using the theories of innovation to predict industry change / Clayton M. Christensen, Scott D. Anthony, Erik A. Roth.
p. cm.
Includes bibliographical references and index.
ISBN 1-59139-185-7
1. Business planning. 2. Business forecasting. 3. Technological innovations—Forecasting. I. Anthony, Scott D. II. Roth, Erik A. III. Title.
HD30.28.C54 2004
658.4′0355—dc22
2004004399

The paper used in this publication meets the requirements of the American National Standard for Permanence of Paper for Publications and Documents in Libraries and Archives Z39.48-1992.

CONTENTS

IN GRATITUDE

The seeds of *Seeing What's Next* were sown almost a decade ago with the publication of the first article describing the phenomenon of disruptive innovation. Readers sensed that the concepts of disruptive innovation were strategically important; but when they asked for guidance in addressing the opportunities and threats that disruption creates, it was hard to offer convincing solutions. Although a few companies had grappled successfully with disruption on occasion, no company had consistently done it well. There just wasn't enough data from the past from which to draw conclusions. Somewhere along the line, I realized that the paralysis I was experiencing had its roots in a flawed paradigm that has guided much of the way we train managers and do research about management—the belief that decisions should be grounded in solid analysis of data. The problem with this paradigm is that when managers attempt to do something that has not been done before, or when the future is going to be different from the past, the paradigm breaks down. Data is only available about the past. I then realized that with data or without it, every time managers take an action, and every time they look into the future, they use a theory to guide their plans and actions—because a theory is a statement of what causes what, and why.

That realization is what led us to write this book. We have attempted to outline a process for using the theories of innovation to predict industry change, even when convincing data is not yet available. We had already written pieces about how disruption might reshape the health care and semiconductor industries and how disruption could be applied at a country level. We further researched these topics and investigated new industries such as telecommunications, education, and aviation to generate a robust way to use theory to predict industry change.

Most of the findings in the book are based on primary and secondary research from sources of the sort that business strategists and analysts would find readily available. We did this to support a key message of this book: The level of one's insight is not strictly a function of the level of one's access to unique or proprietary information. Armed with the right concepts, anyone can produce insights that surpass those of analysts who are knee-deep in industry minutiae.

We apologize to two groups of readers: those from outside the United States and those looking for detailed financial formulas. Most of the book—save for chapter 9—focuses on developments within the United States. In our increasingly global economy, this is clearly an oversimplification. The tools we provide in this book for using the theories of innovation, however, are applicable globally. Strategy and innovation are somewhere between unguided trial-and-error experimentation and rules-based science. Thus the tools in this book are designed to help strategists and planners recognize patterns and interpret the meaning of events as they unfold. Pattern recognition is the best mode of decision making at this point, given the present state of knowledge. Formulas and rules are not yet feasible.

Building comprehensive theory is an iterative process. The lap around the theory-building track in *The Innovator's Dilemma* built the basics of the core theory. The lap around the track in *The Innovator's Solution* detailed important additional theories and began to show how to put the theories into action. This lap continues along the identification and application track—showing how to identify circumstances and understand the implications of those circumstances. The next lap should move toward developing more detailed metrics that indicate circumstances and quantifying the implications of those circumstances.

Listing three authors for this book is quite disingenuous. Dozens of people helped us develop and refine the thoughts in this book. Research partners and former students Sally Aaron, Dan Abassi, Will Clark, Raj De Datta, Carl Johnston, Dr. John Kenagy, Marie Mackey, Michael Raynor,

Nate Redmond, Dr. Chris Robison, Chirag Shah, David Sundahl, and Matt Verlinden each contributed a great deal to the ideas in the book.

We owe a particular debt to the industry and topical experts and friends who helped refine important pieces of *Seeing What's Next*. Jeff Campbell, Pete Cornell, Robert Crandall, John Earnhart, Alexander Edlich, Paula Ford, Lib Gibson, Kevin Goodwin, Joe Graba, Dr. Jerry Grossman, Stuart Hart, Reed Hundt, Laura Ipsen, David Isenberg, Kevin Kennedy, Steven King, Ted Kolderie, Rick Krieger, Gina Lagomarsino, Eric Mankin, Bob Martin, Neale Martin, Joel Meyerson, Steve Milunovich, Bernard Nee, David Neeleman, Stagg Newman, Bob Pepper, Wilfred Pinfold, Michael Putz, Nagi Rao, Rick Rotondo, Chris Rowen, Tim Sloan, Donna Soave, Justin Steinman, Sue Swenson, Tony Ulwick, Jon Wilkins, and Ron Wolk each helped us fine-tune the thoughts presented here.

We would like to thank a number of colleagues at Harvard who were particularly helpful in this effort: professors Liz Armstrong, Dr. Richard Bohmer, Tom Eisenmann, Clark Gilbert, Tarun Khanna, and Steve Spear. My office manager, Christine Gaze, has helped with patient competence in ways too numerous to count. Hollis Heimbouch, whose deft editorial hand has helped to shape this manuscript, has been a wonderful, supportive editor. All of these people have helped build what you are about to read.

I am particularly grateful to my coauthors, Scott Anthony and Erik Roth, and their wives for dedicating a couple of years in the prime of their lives after they finished our M.B.A. program to write this book with me. Whereas my own mind seems to work best with abstractions and concepts, they have demonstrated greater intellectual agility as they helicoptered up and down between exploring high-level concepts and developing practical tools for using them. Over the years in my classroom, I've developed a pretty good eye for capable students. Scott and Erik are two of my very best. They have done most of the work and writing in this book. I am honored to have their names next to mine on the cover.

As with the other books I have written, my wife, Christine, and my children, Matthew and his wife Elizabeth, Ann, Michael, Spencer, and Katie, have patiently helped me bounce these ideas around at home and in the car. I love them and am grateful that I belong to them forever.

—Clayton M. Christensen
Boston, Massachusetts

Harvard Business School Dean Kim Clark likes to refer to the HBS experience as "transformational." I was highly skeptical until I walked into Clayton Christensen's class in the fall of 2000. I had found my transformational experience. Clay is a walking paradox: brilliant but humble; conceptual but practical; an academic with real-world experience; highly successful yet willing to bend over backward to help anyone in need. He struck a chord that changed not only the course of my Harvard experience, but my career trajectory as well. I am eternally grateful that he has provided me the opportunity to help take his ideas in new, exciting directions. Special thanks as well to my other coauthor, Erik Roth. Erik's unique perspective—thoughtful, inquisitive, analytical, and always provocative—helped me look at concepts in valuable new ways. Many important concepts resulted from hours of invigorating debate between Erik, Clay, and me.

Most of the research and writing for *Seeing What's Next* took place while I was a researcher for Clay at the Harvard Business School. Deciding that I wanted to keep wrestling with these ideas, I joined Innosight, a consulting firm founded by Clay. My colleagues at Innosight—Matt Christensen, Tara Donovan, Romney Evans, Matt Eyring, Mark Johnson, and Cheryl Riley—teach me something different each day. They have been very supportive as the work involved in this book continued well past my departure from the Harvard Business School.

A number of friends and family helped along the way. My mother, Bonnie, is the only external person who can claim to have read and reviewed *every single* draft of *Seeing What's Next*. My father, Robert, my brothers, Michael and Peter, and my sisters, Michelle and Tricia, all provided valuable support. My grandfather Robert Anthony—author of close to one hundred books—served as an inimitable example to emulate.

This has not been an easy course. Patience has never been one of my virtues, and any author who is not blessed with the natural ability to write perfectly (as I, sadly, am not) will testify that the one thing writing requires is patience. With me every step of the way, as always, has been my wife, Joanne. During the course of the book's development, we got engaged. Between drafts three and four, we got married. Without her support, I am quite sure this book never would have been finished. Her friendship, compassion, loyalty, determination, and love have guided me on the most important and most rewarding transformational experience of my life. I feel lucky just to know her, let alone have the privilege of spending the rest of my life with her. So I dedicate this work, as I dedicate all things, to her.

—Scott D. Anthony
Boston, Massachusetts

This experience with all of its ups and downs and its twists and turns has been one of the most rewarding of my professional life, and most of all I am indebted to Clay Christensen. None of us would have guessed that a month-long consulting project following our M.B.A. coursework would have evolved into a book three years later, but Clay provided the vision and energy to make this possible. His willingness to invite students to participate in this endeavor and his faith in our abilities reflect his dedication to teaching, his generosity, and his patience. Thank you also to Scott, my other coauthor and comrade, for three years of stimulating dialogue, vigorous debates, lots of laughs, and for driving this project across the finish line.

Furthermore, I owe a debt of gratitude to my wife and soul mate, Kate, whose love and support continue to inspire me. This has been a long process and she has been a source of limitless encouragement. And there is my other girl, Annabel, whose boundless energy exposed me to the distinctive beauty and quiet of morning twilight, teaching me that I can be productive and creative at all hours.

Last, I must thank my extended family of supporters: my parents, Richard and Leslie Roth; Richard and Nan Ruben; and the uncles, Greg, Nick, Erik, and Gregory, for always being there as advisers, editors, cheerleaders, and babysitters, among countless other roles.

—Erik A. Roth
Cambridge, Massachusetts

INTRODUCTION

Imagine the year is 1876. You work for a leading bank. Your boss comes to you with a deceptively simple question: "Alexander Graham Bell has developed a way to transmit voice over a wire. What does this mean?" How would you go about answering that question? The world's leading communications company, Western Union, called Bell's innovation a "toy." No data exists to help guide your analysis.

Step forward about one hundred years to 1978. You work for a consulting firm. Your team manager comes to you with a deceptively simple question: "AT&T is testing a mobile phone service. What does this mean?" How would you go about answering that question? How could you interpret the subsequent choices companies made as they commercialized this new technology?

Finally, move forward to 2004. Imagine you work for a leading telecommunications service provider. Your boss asks a deceptively simple question: "I've read everywhere about this explosion of local high-speed wireless data networks using a technology called 802.11. What does this mean?" How would you go about answering that question? If you were an investor, how could you interpret the choices the company makes to either commercialize or ignore the technology? Is the company being wise or myopic?

Day in and day out, millions of us take action based on what we think the future will hold. Investors buy shares when they think a company's future is bright and sell shares when they think a company's future is dim. Analysts try to understand what the future holds so they can dispense insightful nuggets of wisdom to their clients. High-level executives try to separate signal from noise, threats from opportunities, and act accordingly.

All of these people observe actions taken by managers and try to determine the impact those actions will have on the future. Typically, the biggest questions on their minds are: How will this innovation change an industry, and what impact does this have on the firms I care about?

The questions affect every industry. Consider aviation. Even before the terrorist attacks of September 11, 2001, everyone considered it a dismal industry. Using Harvard Business School professor Michael Porter's five forces framework—looking at competitors, potential entrants, buyers, substitutes, and suppliers—can help explain why this is so today.[1] But will it always be a dismal industry? What do we make of the glimmers of hope such as Southwest? Will JetBlue continue to prosper? Which are real opportunities and which are transient? Can discount airlines, regional jets, or even point-to-point air taxis reshape aviation?

Think about semiconductors. Intel has a long history of success based on producing ever higher performing chips. What would signal that the game is changing, meaning what was successful in the past would no longer guarantee success in the future? What implication would that change have on the industry's value chain?

Or think about health care. Almost every newspaper story about health care complains about spiraling costs and growing consumer dissatisfaction. Could this possibly be *good* news? What kinds of innovations could heal health care?

Seeing What's Next shows how to use the theories of innovation developed in *The Innovator's Dilemma*[2] and *The Innovator's Solution*[3]—and introduces some new ones as well—to answer these sorts of questions.

The Innovator's Dilemma developed a theory that explained why launching new-growth businesses is so hard; *The Innovator's Solution* showed would-be innovators how to use theory to make the process of launching growth businesses more predictable. Both books were "inside-out," written from the perspective of enterprise-level decision makers charged with framing and executing strategy.

Seeing What's Next shows how to use these theories to conduct an "outside-in" analysis of how innovation will change an industry. The book demonstrates that a structured, rigorous process of looking at an industry or industry segment through the lenses of the theories of innovation provides powerful insights not readily observable to the untrained eye. It will help senior executives, strategists, industry analysts, investors, and all others who need to make decisions or recommendations based on their assessment of an industry's future.

Before diving in, let us step back and reintroduce the core theories behind this work, explain why theory is an appropriate tool for predicting industry change, and explain how we organized the book to show how to see what's next.

The Core Theories of Innovation

Good management theory provides situation-specific statements of cause and effect.[4] Good theory has two components:

1. An underpinning of a robust circumstance-based categorization scheme that provides a guide to the situations managers encounter
2. A causal statement that explains *why* certain actions lead to certain results and that describes how the result of actions will vary from one circumstance to the next

At the core of *The Innovator's Dilemma* and *The Innovator's Solution* are three important theories that untangle the messy process of innovation: the disruptive innovation theory; the resources, processes, and values theory; and the value chain evolution theory. Let's briefly revisit each theory. Seasoned readers can skip ahead to the next section.

The Disruptive Innovation Theory: Simple, Cheap, Revolutionary

The disruptive innovation theory points to situations in which new organizations can use relatively simple, convenient, low-cost innovations to create growth and triumph over powerful incumbents.[5] The theory holds that existing companies have a high probability of beating entrant attackers when the contest is about *sustaining* innovations. But established companies almost always lose to attackers armed with *disruptive* innovations.

FIGURE I-1

The Disruptive Innovation Theory

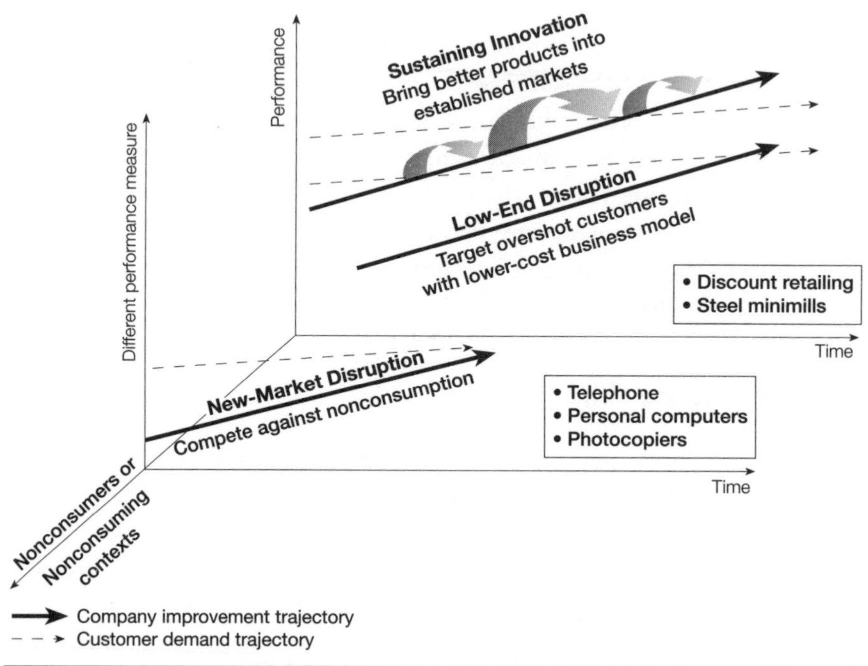

Figure I-1 illustrates the disruptive innovation theory. On the figure are two types of improvement trajectories. The solid lines illustrate company improvement trajectories. They show how products and services get better over time. The dotted lines show customer demand trajectories—not the quantity customers demand, but the performance they can use. As these trajectories suggest, a customer's needs in a given market application tend to be relatively stable over time. The figure shows three types of innovations: sustaining innovations, low-end disruptive innovations, and new-market disruptive innovations.

Sustaining innovations, illustrated by the curved arrows, are what move companies along established improvement trajectories. They are improvements to existing products on dimensions historically valued by customers. Airplanes that fly farther, computers that process faster, cellular phone batteries that last longer, and televisions with incrementally or dramatically clearer images are all sustaining innovations.

Disruptive innovations introduce a new value proposition. They either create new markets or reshape existing markets. There are two types of disruptive innovations: low-end and new-market. Low-end disruptive innovations can occur when existing products and services are "too good" and hence overpriced relative to the value existing customers can use. Nucor's steel minimill, Wal-Mart's discount retail store, Vanguard's index mutual funds, and Dell's direct-to-customer business model were all low-end disruptive innovations. They all began by offering existing customers a low-priced, relatively straightforward product.

The second type, new-market disruptive innovations, can occur when characteristics of existing products limit the number of potential consumers or force consumption to take place in inconvenient, centralized settings. The Kodak camera, Bell telephone, Sony transistor radio, Xerox photocopier, Apple personal computer, and eBay online marketplace were all new-market disruptive innovations. They all created new growth by making it easier for people to do something that historically required deep expertise or great wealth. Figure I-1 shows how new-market disruptions bring consumption to "nonconsumers" or "nonconsuming contexts."

The Resources, Processes, and Values Theory: The Building Blocks of Capabilities

The resources, processes, and values (RPV) theory explains why existing companies tend to have such difficulty grappling with disruptive innovations.[6] The RPV theory holds that resources (what a firm has), processes (how a firm does its work), and values (what a firm wants to do) collectively define an organization's strengths as well as its weaknesses and blind spots.

Resources are things or assets that organizations can buy or sell, build or destroy. Processes are the established patterns of work by which companies transform inputs into outputs—products or services—of greater worth. Values determine the criteria by which organizations allocate their resources. Figure I-2 lists the components of these three factors.

Consider Microsoft. Microsoft's resources include its more than 50,000 employees, thousands of software coders, stable full of developed products, billions of dollars in cash, and collection of strong brands and products, including its popular Windows operating system. Microsoft's critical processes include those by which it conducts software development, market research, budgeting and planning, and product

FIGURE I-2

Resources, Processes, and Values

Resources	Processes	Values
Things or assets that organizations can buy or sell, build or destroy. Examples: • People • Technology • Products • Equipment • Information • Cash • Brand • Distribution channels	Established ways companies turn resources into products or services. Examples: • Hiring and training • Product development • Manufacturing • Planning and budgeting • Market research • Resource allocation	The criteria by which prioritization decisions are made. Examples: • Cost structure • Income statement • Customer demands • Size of opportunity • Ethics

distribution. The company's prioritization criteria typically lead executives to allocate resources to opportunities whose revenues are large enough and whose gross margins are attractive enough to sustain or improve the company's overall financial performance.

The RPV theory argues that organizations successfully tackle opportunities when they have the resources to succeed, when their processes facilitate what needs to get done, and when their values allow them to give adequate priority to that particular opportunity in the face of all other demands that compete for the company's resources. Incumbent firms master sustaining innovations because their values prioritize them, and their processes and resources are designed to tackle precisely those types of innovations. Incumbent firms fail in the face of disruptive innovations because their values *will not* prioritize disruptive innovations, and the firm's existing processes do not help them get done what they need to get done. For example, Microsoft has struggled to respond to the Linux operating system. Why? It isn't a resource problem. Microsoft would have to design new processes to facilitate the design of modular-architecture software, but this is a solvable problem. But it is *very* difficult for Microsoft to prioritize a Linux-based business compared to the other investment opportunities that promise profits that are more attractive. Companies offering Linux make money in *very* different ways than Microsoft does.

The Value Chain Evolution Theory: Integrating to Improve What Is Not Good Enough

Producing a product or delivering a service requires completing a set of activities. Companies have a choice: They can choose to integrate, executing most of the activities themselves, or they can choose to specialize and focus on a narrow range of activities, relying on suppliers and partners to provide other elements of value added. The final core theory of innovation—the value chain evolution (VCE) theory—assesses whether a company has made the right organizational design decisions to compete successfully.[7]

On the surface, the VCE theory is breathtakingly simple. The theory suggests companies ought to control any activity or combination of activities within the value chain that drive performance along dimensions that matter most to customers. Directly controlling, or integrating, an activity gives companies the ability to run experiments and push the frontier of what is possible. Integration gives firms a full platform to run experiments to solve problems caused by unpredictable "interdependencies" between activities. These same interdependencies can frustrate specialist firms that try to focus on a single piece of a product's or service's value chain. When a specialist's piece interacts unpredictably with components that other companies design and make, it typically results in poorly performing, unreliable products.

Consider IBM's early mainframe computers. IBM needed to improve the mainframe's overall performance. It integrated the design and assembly processes for individual components and the entire computer. Complete control gave IBM the design freedom to experiment and improve mainframes to meet customer needs. A modular, nonintegrated strategy would have produced an underperforming product that customers would have rejected.[8] Similarly, current handheld wireless device performance falls short of customer needs along dimensions such as battery life. By controlling the entire product architecture for its Blackberry personal communications device, Research in Motion (RIM) developed a battery that lasts three weeks between charges. This innovation helped RIM become a market leader.

The performance improvements that integration provides come at a cost, however. Integrated architectures tend to be relatively inflexible. Integrated companies tend to react relatively slowly. Therefore, the theory suggests that companies ought to outsource activities that don't influence the characteristics of a product or service that customers deem (or will deem) most critical. Specialists can better optimize those pieces of the value chain.

Modular architectures that facilitate (or permit) disintegration sacrifice raw performance in the name of speed to market, responsiveness, and convenience. This sacrifice allows companies to customize their products by upgrading individual subsystems without having to redesign an entire product. They can mix and match components from best-of-breed suppliers to respond conveniently to individual customers' needs.

Dell introduced convenience and customization to the personal computer market. It did this by tightly integrating across the key interfaces in the supply chain, integrating across the interface with the customer, and outsourcing component design and production to specialist providers. It followed the VCE theory's golden rule: Integrate to improve what is "not good enough" (speed, customization, and convenience) and outsource what is "more than good enough" (the computer's architectural design).

Solving the hard problems allows firms to capture value. Forward-thinking firms move to solve tomorrow's hard problems, because solving tomorrow's hard problems creates tomorrow's profits. They unwittingly follow the advice of hockey legend Wayne Gretzky who, when asked what made him so great, replied that he always tried to skate to where the puck was going to be, not to where it was.

The Power of Good Theory

The only way to look into the future is to use these sorts of theories, because conclusive data is only available about the past. Many people who make their livings providing advice about the future express deep skepticism about theory-based prognostication. Interestingly, these people are *already* theory-based analysts. They just don't know it. And often, the theory they use is poor.

Consider analysts working for a large investment bank. How do they predict industry change? Typically, they gather historical data, determine trends, and make projections. They extrapolate a firm's past earnings to determine its future cash flow and then discount that cash flow at a risk-adjusted rate to determine firm value. Whether a company met projected earnings in the recent past heavily influences the analysts' forecasts of future stock price performance. In doing so, they rely on an implicit theory: The past is a good predictor of the future.

Now consider management consultants, seeking to tell a company how to organize its sales force. Many of them approach this challenge by identifying a "best-practice" company and gathering gigabytes of data "proving" how the company's particular sales force design is key to its success. If their client would only imitate that comparison company, the consultants say, they too would reap the rewards. Again, the consultants base their recommendation on an implicit assumption: Companies find success when they mimic actions taken by successful or "excellent" companies.

Sometimes these assumptions are correct and lead to great insight. But sometimes they don't. The past is a good predictor of the future only when conditions in the future resemble conditions in the past. And what works for a firm in one context might not work for another firm in a different context.

Furthermore, those who require data or best-practice comparison companies to make decisions about the future must throw their collective hands up when unassailably conclusive quantitative data doesn't exist. And the truth is, data only becomes conclusive when it is too late to take action based on its conclusions. For example, consider a senior management team that becomes aware of a new start-up company in its industry that is employing a radically different business model. Can the team safely dismiss the idea as an unrealistic entrepreneur's dream? Or does it portend the beginning of an important change in the industry's future? Is this signal, or is it noise? How can they tell? By the time the evidence is clear, it will be too late to take action.

The best way to make accurate sense of the present, and the best way to look into the future, is through the lens of theory. Good theory provides a robust way to understand important developments, even when data is limited. And theory is even *more* helpful when there is an abundance of data. This is the critical challenge of the Information Age. With more information available, it is harder to discern what information really matters. Theory helps to block out the noise and to amplify the signal.[9]

How Theory Provides Insight into the Past: Two Case Studies

Before we show how to use theory to see the future, let us step back and show how the core theories in *The Innovator's Dilemma* and *The*

Innovator's Solution illuminate two seminal developments in the U.S. telecommunications industry: the birth of the telephone and the explosion of wireless technologies (see "Telecommunications Overview" for a brief primer on the U.S. telecommunications industry).

Specifically, this section shows how theory explains why the telephone led to the demise of the day's leading incumbent communications company—Western Union—and why wireless technologies are *not* leading to the demise of today's leading incumbents. In short, the disruptive innovation and RPV theories tell us why. The telephone was disruptive relative to Western Union's telegraphy service. Western Union had the resources to master the telephone, but its values led it to focus on its existing business to its ultimate peril. In contrast, although wireless telephony *could* have been disruptive, today's incumbent telephony companies co-opted wireless technologies in a way that sustained their existing business model.

The past might seem a strange departure point for a book that explains how to see the future. We start with history, however, to show that theory can help explain why things happened the way they did. After all, we cannot hope to use something to explain the future if it cannot explain the past.

The Birth of the Telephone

Alexander Graham Bell did not invent the technology that would become the telephone with the intention of toppling Western Union.[10] He set out to help Western Union improve its core telegraphy business. Bell offered the patents he had received on his technology to Western Union for a mere $100,000, roughly $1.7 million in today's dollars.[11]

Western Union turned Bell down. One way to interpret this mistake—and the way people typically explain such mistakes—is to point to management myopia. Rapidly growing and attractively profitable, Western Union simply made a colossal mistake.

In declining to purchase the innovation, Western Union president William Orton famously said, "What use could this company make of an electrical toy?"[12] Spurned by Western Union, Bell and his backers chose to commercialize the technology with a licensing business model. The first telephone company appeared in New Haven, Connecticut, in 1878. Although the technology could only carry the telephone signal for a few miles at the outset, a new market emerged. Americans used the

Telecommunications Overview

Because we will use a number of telecommunications case studies in this book, we thought it would be helpful to provide a very brief overview of the seminal events over the past seventy years that have shaped the telecommunications context in the United States.

In 1934, the government passed the Communications Act, which put in place the regulatory scheme that still largely governs the telecommunications industry today. The act envisioned a monopoly telephone service company and created the Federal Communications Commission (FCC) as its primary regulator. The act, passed a half century after telephony began and after more than 30 percent of all households had a telephone, was the first time the government recognized the telephone as a "public good" requiring government regulation to protect the public interest.

The telecommunications system was an end-to-end monopoly until the 1960s. The first cracks appeared in the customer premise equipment (CPE) market. Then long distance providers—most notably MCI—entered to further the assault.

The breakup of AT&T in the early 1980s solidified MCI's status as a true competitor to AT&T. The Modification to the Final Judgment in 1982 split AT&T into a long-distance company and a group of twenty-two Regional Bell Operating Companies, known as RBOCs or Baby Bells. AT&T was allowed to enter into new business lines. The RBOCs would continue to be regulated entities because they controlled the golden "last mile," the set of wires that connect individual homes and businesses to the rest of the network. A series of mergers between the RBOCs resulted in the creation of four large local telephony companies: BellSouth, SBC, Verizon, and Qwest.

Buoyed by the success of deregulatory efforts in encouraging competition (with the consequent improvements in pricing and performance), the government turned in the 1990s to what it perceived to be the last refuge of the old AT&T monopoly—the RBOCs. The complicated Telecommunications Act of 1996 (Telecommunications Reform Act or TRA) chipped away some of the last remaining barriers to competition in the local market by forcing the RBOCs to share their lines with competitive local exchange carriers (CLECs). Many CLECs failed, some quite spectacularly, but competition in local markets did begin to increase. We will use our theories to explain why these things happened in the way that they did in this and subsequent chapters.

telephone as a more convenient way to communicate with people living nearby.

Bell licensees in geographically exclusive areas emerged to provide simple, point-to-point services. In return, they gave Bell a portion of the revenue stream. The first adopters were local businesses that saw the technology as a way to improve communication within offices and between nearby offices. The number of subscribers grew quickly as wealthy households found the telephone to be a useful way to communicate with their servants. By 1879, users had purchased more than 17,000 phones. By 1900, the number of users swelled past 1 million.

At first, the telephone's growth had little impact on Western Union's core business because telephone wires could only carry local calls. By 1900, only 3 percent of the average number of daily calls were long distance. But even then, it was clear Western Union had made a colossal mistake. That year, Western Union reported an annual net income of roughly $6 million. The Bell telephone companies, reorganized underneath a single corporate entity called the American Telephone and Telegraph Corporation (AT&T), reported net income of more than $13 million.[13] By 1910, the challenger consumed the incumbent, as AT&T acquired a controlling interest in Western Union. Although the government eventually forced AT&T to divest its interest in Western Union, AT&T went on to become one of the largest, most powerful, and most profitable companies in the world—all based on a technology that Western Union didn't think enough of to purchase for $100,000.

The popular press might recount the story in this manner, probing no deeper than the level of "management mistakes." However, something doesn't sit right.

Was Western Union's management incompetent? Consider these facts. Western Union's management somehow was smart enough to create what historians consider the "first nationwide multiunit modern business enterprise."[14] Yes, they did initially discount the telephone. But *so did the telephone's inventor*: "Initially, [Bell] presented the telephone as a novelty rather than a communications device capable of challenging Western Union."[15] In fact, Bell even patented his mechanism to use electricity to send the human voice across a wire under the name "Improvements in Telegraphy."[16]

A more upbeat assessment grants the intelligence of the management team and curses the randomness of the process of innovation and competition. There was just no way that Western Union could have anticipated that the telephone would ever get good enough to be any kind of competitive threat. As one historian notes, "Neither telephone nor telegraph industry leaders could conceive of a world in which ordinary people would pick up a telephone just to chat with friends and relatives."[17]

And what is more, Western Union *didn't* miss the signs indicating the telephone was important. In fact, it aggressively attempted to enter the telephone industry. It hired another well-known inventor, Thomas Alva Edison, to develop a competing system. But Western Union's heart wasn't in the fight. It decided to cede the local telecommunications market to the Bell operators.[18] In return, the Bell companies agreed to give Western Union a royalty on their activities, to pass long-distance telegraph communications to Western Union, and to not enter Western Union's lucrative data communications market. Western Union's failure was not to ignore the telephone, but to decide to focus its efforts on its highly profitable core business.

So why did Western Union make decisions that history would deem shortsighted? Theory suggests a four-part answer:

1. ***The telephone was a new-market disruptive innovation.*** The telephone was a classic new-market disruptive innovation relative to the telegraph. Because it could only carry a signal for a few miles, it could not be used in the market that existed at the time—long-distance communications. It facilitated people's ability to communicate conveniently over short distances. It did so using the familiar nature of speech. Users did not have to learn the specific skills required of telegraph operators. This led users to happily purchase a seemingly limited innovation; the limited innovation was a lot better than walking a mile to talk with someone. Early growth provided funds for further investment and improvement.
2. ***Western Union's resources, processes, and values meant that what ultimately became the right course appeared to be unattractive at the outset.*** Western Union passed on the telephone because it sensibly prioritized investing in its core market. The stream of investment decisions and Western Union's overwhelming focus on long-distance telecommunications reflected the fact that Western Union's core business was a very attractive money-

making proposition at the time. Its long lines carried important data that provided updates on financial markets and helped railroads manage their operations. Telegraphy had made Western Union a great company. That was where its best-paying customers were. In all new-market disruptions, incumbents have substantial scope to profitably improve the performance of their products and services. Western Union's best long-distance data communications customers—railroads, newspapers, and financial brokerage houses—needed still better long-distance data and communications services. They were seemingly uninterested in the ability to chat with colleagues.

3. ***Western Union saw entrants improving. However, investments in the core business kept trumping investments in the new business.*** The telephone market was just too small to materially affect Western Union's financial position. What looked huge to hungry upstarts did not meet the company's growth needs. Even though Western Union had the resources to succeed, it didn't make sense for it to waste time building a new market when there was so much more growth to be gleaned from its existing customers—because the banks and railroads were growing rapidly too.

   Once the telephone had taken root in local communications, the Bells' motivation to make more money gave them an incentive to overcome the technological hurdles limiting the number of users the system could effectively handle and the distance calls could travel. Solving these problems would allow the Bell companies to better serve their current customers and attract new customers. Sustaining innovations such as the switch and the loading coil (discussed in chapter 1) propelled the telephone up its technological improvement trajectory, ultimately allowing telephone companies to offer a legitimate long-distance service within fifteen to twenty years of the telephone's invention.

4. ***By the time the right course was clear, it was too late.*** The skills entrants accumulated during their up-market march essentially barricaded Western Union from the market. Telephone companies honed the unique ability to deliver voice telephony service. By the time the service was good enough to begin affecting Western Union materially, the company lacked the ability to

> respond. Entering companies had sharpened their ability to oversee telephony networks with their complicated web of interconnections, manage voice traffic, and market services to a far-flung group of consumers. Many of the skills Western Union honed in the telegraphy world would not help it here. It would be as difficult for Western Union to beat back the telephone in the early 1900s as it would have been for telephone companies to take on the telegraph in the 1870s.

In short, Western Union passed on the telephone for predictable reasons. The telephone grew, predictably. And Western Union predictably couldn't respond. Success wasn't due to good management; failure wasn't due to bad management. Each management team made the appropriate profit-optimizing decisions given the starting positions in which the firms found themselves.

This case study demonstrates how it is incorrect to assume that *information* itself will cause managers to act or react in a consistent, predictable way. We need also to understand the context in which the managers are operating—because the context gives meaning to the information. The theories of innovation help to understand the forces that shape this context and influence natural decisions. Theory illuminates the signals indicating important developments and explains the likely implications of these developments on industry participants.

The Explosion of Wireless Technologies

If you are reading this in an airport lounge, glance to your left. Now glance to your right. Odds are that half of those you see are chatting away on a mobile telephone. By 2004, more than half the people in the United States carried a wireless phone. Penetration was higher in many countries. Indeed, the last twenty-five years have seen an explosion in wireless communications.

Cellular voice service arrived in the 1980s with the introduction of the bulky car phone. From a technological point of view, this was a disruptive innovation. Wireless voice's basic functionality as measured by call quality was relatively worse than the wired alternative. Call quality was poor, battery life was inadequate, and the phones were bulky and expensive. Early users, however, valued the technology for its inherent attributes of additional convenience—the ability to make calls in a new context, anywhere and anytime.

As with all disruptions, the pursuit of growth and profit caused wireless phones to improve quickly. Within fifteen to twenty years, by the late 1990s, signs began to emerge that people were "cutting the cord" and ditching their wired phones entirely. Public pay phone use began to drop precipitously in 1996.[19] College students, young adults sharing apartments, and parents seeking a second line for their chatty teenagers all began choosing mobile telephones over a fixed-line connection. Long-distance providers next felt the pressure. Wireless companies bundled increasing amounts of "anywhere" and "anytime" minutes that included ostensibly "free" long distance. This put tremendous pricing pressure on the long-distance market, as many users would simply wait until their "anytime" minutes kicked in and then use their mobile phones to make long-distance calls. By 2002, analysts estimated that overall wireless phone usage had already displaced as much as 26 percent of landline minute usage. By 2003, up to 7.5 million people had "cut the cord."[20]

At first blush, something seems strange. Wireless technologies feel like new-market disruptions. They enable consumption in a new context. They initially underperformed on metrics such as call quality that were highly valued in the old context but brought new benefits such as convenience. We expect new firms to drive disruptive innovations, but incumbents suspiciously litter the list of the leading wireless players. Verizon, SBC Communications and BellSouth (through their joint ownership of Cingular Wireless), AT&T, Sprint, Bell Canada Enterprises, Deutsche Telekom, and NTT all have significant wireless positions. How did incumbent telecommunications firms succeed at disruption whereas Western Union failed at it?

Perhaps managers grew smarter over the course of a century. However, at the same time these telecommunications companies were successfully grappling with disruption, disruptive innovations were pummeling other historically well-regarded companies such as Compaq, Digital Equipment Corporation, General Motors, IBM, Sears, and U.S. Steel. Although it is possible that smarter managers ended up in telecommunications firms, it seems more likely that something else was going on.

Indeed, theory highlights the critical choices that ultimately determine an innovation's fate. Wireless voice is a disruptive application vis-à-vis wired voice. Entrants such as McCaw Cellular indeed created high levels of growth.[21] But four critical decisions—two by the government and two by industry players—put wireless technologies on a path where incumbents had both the skills and the motivation to co-opt and ultimately master the innovation. Let's go through each choice:

1. ***The government granted licenses to incumbents, starting them down a complementary path.*** The government played an important role in wireless's development. In 1981, the FCC granted one license (the so-called "B" block) to the established telephone companies and one license (the so-called "A" block) to a nonwired-line player. From the start, incumbents not surprisingly tried to build a system that complemented their asset bases and business models. For their part, many entrants decided to build their businesses with the intent of selling them to incumbent providers (discussed below). They naturally designed their businesses to be as compatible with wireline telephony as possible.
2. ***Industry players designed a service to reach the most demanding customers.*** Most service providers imagined their best hope for success was to sell to mobile professionals with a high willingness to pay for mobile services.[22] Satisfying demanding business customers required creating networks capable of delivering reliable, uninterrupted, and continuous service in a moving vehicle while minimizing the capital cost of building cellular towers. This was a tall order. It influenced every subsequent design choice. Phone manufacturers compensated for sparse towers by producing expensive, high-power handsets capable of picking up the faintest signals. Cellular coverage focused on major roadways. Highly mobile users moved around the country faster than local providers could build infrastructure or secure the rights to offer service in all markets. The solution? Complex roaming agreements.

   These development choices minimized the conflict between wired and wireless businesses. Wireless followed a path that was highly complementary and additive. Building networks allowed established companies to earn more money from their best customers (and even get into new markets). The basics of operating a wireless business mirrored the basics of operating a wired business. Wireless demanded amortizing high up-front capital expenditures over a large subscriber base. As the industry developed, wireless carriers experimented with innovative bundles and service plans. But the minutes of use models between wireless and wired line providers remained very similar. Both types of providers drove as much usage as possible over existing assets. The average wireless minute had high gross margins that were similar to margins on the average residential wired-line voice minute. These similar business models motivated incumbents to go after wireless opportunities.

3. ***Entrants built overlapping delivery networks that made incumbent co-option natural and easy.*** Tightly overlapping delivery networks enhanced incumbent firms' ability to master wireless technologies. In the late nineteenth century, the original telephone companies largely built their own separate networks. Nascent wireless companies, on the other hand, naturally wanted their customers to have the ability to exchange calls with landline users. These calls *had* to travel over existing networks. Even today, unless you place a call to a subscriber who is on your network (for example, Sprint, Cingular, and so on), chances are your call uses the local telephone company's landlines.[23]

   Relying on these networks made sense: It reduced development costs and provided a way out of the chicken-and-egg problem of building large networks with the expectation that customers would come. But interconnecting with the local phone company meant meeting its specifications and conforming to its billing mechanisms. It required a business model that made sense to participants throughout the value network. Even service providers who ignored wireless in the industry's early days could acquire firms once wireless got too big to ignore, so that they could continue to benefit from the ongoing growth.
4. ***The government forced incumbents to set up separate subsidiaries.*** The government unintentionally aided incumbents by mandating that existing telephone operators set up separate business units to commercialize the licenses. That decision made it easier for incumbents to avoid the internal battles that often make it difficult for incumbents to respond to disruptive innova-

tions. Separate divisions have always been a viable way for incumbents to deflect disruptive entrants.

In short, a series of decisions facilitated incumbent co-option of wireless technologies (even this co-opted path presents future challenges, discussed in chapter 10). These decisions meant the incumbents had resources to deploy in attacking the opportunity, processes that facilitated their ability to build and maintain wireless networks, and values that motivated them to pursue wireless opportunities.

What path could companies bent on disruption have followed? They would need to have targeted less demanding customers, such as parents who wanted to keep in touch with their children within a local neighborhood. They would need to have built and designed networks specifically to *not* interact with the existing telephone network, limiting calls to customers only. Such a path could have ultimately led to disruption. (We will discuss these sorts of choices more in chapter 3.)

How to See What's Next

The historian in all of us cares about the past. The decision maker in all of us cares about the future. The good news is that theory that explains the past can—if used properly—provide insight into the future. We are not alone in our belief that the theories of innovation can help predict industry change. After *The Innovator's Dilemma*'s release, a number of readers sought to use the principles of disruptive innovation to better understand what was happening in industries that mattered to them. Analysts, investors, and managers from industries as varied as health care, telecommunications, defense, education, and semiconductors have used the concepts to interpret data about what is happening in the present and to understand what the future holds. They found great clarity by looking at their industries through the lenses of the theories presented in *The Innovator's Dilemma* and *The Innovator's Solution*.

Working with many of those who were using our theories raised an issue. Even people who deeply understood the theories struggled to use them in a repeatable and methodical fashion. They didn't know the questions to ask to determine unambiguously before the fact whether an innovation would have a sustaining or disruptive impact on their industry. They didn't know exactly where to look to see developments that mattered, or how to recognize them when they saw them. They didn't know exactly what to look for. Indeed, although many of the ideas embedded in the theories of innovation seem simple, using the theories as a predictive tool can be difficult. To assist in these efforts, we set out to develop and communicate a structured way to use the theories of innovation to look more accurately into the future of companies, industries, and technologies. *Seeing What's Next* reflects the results of this effort.

The analytical tools in this book build on the theories presented in *The Innovator's Dilemma* and *The Innovator's Solution.* Indeed, although *Seeing What's Next* introduces new concepts and frameworks—such as the motivation/ability framework for assessing nonmarket forces—this is not a book about building theory. This is a book about using theory to gain insight into the future. As such, "disruption experts" who understand the core theories in great depth will gain the most from *Seeing What's Next.* But detailed theoretical knowledge is not a prerequisite to use the approach laid out in this book. The numerous lists, tables, and figures aim to make *Seeing What's Next* a practical guide for novices and experts alike. An appendix at the end of the book provides overviews of all of the theories described in the book, and a glossary defines every word to which we ascribe unique meaning. We hope *Seeing What's Next* will help readers to develop an intuition for how to use the theories of innovation to predict industry change.

We begin by presenting a three-part process (summarized in figure I-3) we have developed to use theory to predict industry change. Each of the first three chapters continues to draw on telecommunications examples (see "Why Telecommunications?") to build one part of the process.

Chapter 1 explains how to identify *signals of change*, which indicate situations where change is possible, where we can expect the future to be materially different from the past. In these situations, we can expect firms to emerge with products, services, or business models that look very different from what we have seen in the past. The "entrance" of these firms might be invisible to even the most astute industry watcher because they can incubate seemingly far away from existing markets or seem too inconsequential to matter. But if you know where to look and what to look for, you can spot industry-changing firms before they emerge.

Chapter 2 explains how to evaluate *competitive battles*, head-to-head battles between companies loosely classified as "attackers" and "incumbents." As we will discuss, the process of innovation typically

FIGURE I-3

Process to Predict Industry Change

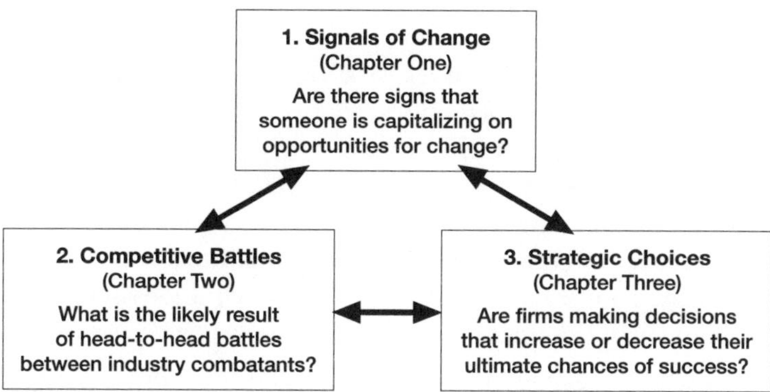

Why Telecommunications?

This book is about using theory to analyze any industry. Much of chapters 1 through 4 and chapter 10 focus on the telecommunications industry. Honing in on a single industry allows us to provide detailed case studies whose purpose is to help our readers begin to build deep intuition about how to use the theories of innovation to understand the past and present and predict the future.

We could have chosen any industry for our deep dive. Telecommunications piqued our interest because it illustrates the pattern that afflicted many high technology sectors from 1997 to 2003. First, rampant hype by the news media and investors inflated the value of most ventures that were new and technology intensive. The bubble burst. Everyone proclaimed that the future just wasn't going to materialize in the way people thought. Hype turned to despair. Now, it turns out that some technologies and companies spawned during the bubble have an exciting future. Could we have predicted this *ex ante*? To us, the telecommunications industry hungered for a way to understand why some innovations succeed while others fail, why some firms thrive while others wither.

(continued)

(continued)

Three additional things about telecommunications piqued our interest:

1. ***It is a large and important industry.*** In the United States, telecommunications-related spending accounts for roughly 3 percent of the gross domestic product (GDP), with the average company spending a growing portion of its revenues on communications and data networking–related expenses.* Coupled with simultaneous advances in computing, telecommunications is one of the drivers of the convergence of multiple high technology sectors. It is a key enabler of the Information Age.

2. ***It has a long and illustrious history of innovation.*** Over the last one hundred years, Bell Labs, the AT&T-funded and wholly owned research laboratory (spun off as part of Lucent Technologies), is credited with countless groundbreaking innovations such as the transistor, the laser, stereophonic and motion picture sound, cellular telephones, and high-definition television. With this long history of important innovations, telecommunications seems like a perfect environment to study the forces of innovation.

3. ***The role of the government:*** Telecommunications presented us with an opportunity to apply the theories of innovation within an industry characterized by heavily regulated competition. Ironically, few people have studied the specific relationship between government intervention and innovation in the telecommunications industry or a more general setting.

*The telecommunications industry, of course, experienced a dramatic decline in the first part of the twenty-first century. FCC chairman Michael Powell told a Senate commerce hearing on July 30, 2002: "This is an industry where nearly 500,000 people in the United States alone have lost their jobs, and approximately $2 trillion of market value has been lost in the last two years." Still, according to the Bureau of Economic Analysis, the telephone and telegraph industries contributed about $265 billion to the United States's $9.2 trillion GDP in 2001 (figures in 1996 terms), or 2.9 percent of GDP. The industries grew at a compound rate of about 7 percent from 1987 to 2001, more than twice as fast as GDP. Information is available at <http://www.bea.gov/bea/dn2/gpox.htm> (accessed 12 February 2003).

leads to entrants intruding on the turf of powerful incumbent firms. Theory provides a way to understand who is likely to emerge triumphant.

Chapter 3 looks at the *strategic choices* that can influence the outcome of competitive battles. We show what attackers can do to tilt the balance of power in their favor and what incumbents can do to withstand attacks.

Chapter 4 rounds out the analytical part of *Seeing What's Next* by looking at the important yet understudied relationship between innovation and nonmarket forces such as government regulation. Understanding the interplay between these forces is critically important to seeing the future of industries that feature heavy government involvement, such as health care, financial services, and education.

The book's next part analyzes six very different topics—the education, aviation, semiconductor, health care, and telecommunications industries, as well as innovation overseas—to show the theories of innovation in action. Each industry-specific chapter uses our approach to show how theory helps to explain why things in the past happened as they did and what is likely to happen in the future. In selecting a broad group of industries, we purposely limit the depth of our analysis on any one topic to show the wide applicability of our approach.

Theory provides important insights into each topic. Chapter 5 shows how disruptive forces could reshape both postsecondary and primary education by bringing education into new contexts. Chapter 6 focuses on how major airlines are in a no-win business because disruptive attackers nip at their heels but they are unable to leave low-margin customers behind in search of more lucrative customers. Chapter 7 suggests semiconductor firms relentlessly pursuing "Moore's Law" are in danger of overshooting increasing swaths of the market, creating opportunities for firms to offer more customized, simpler products. Chapter 8 shows how the theories of innovation see untapped opportunities for disruptive innovations to move treatment from the hospital to the home, from the surgeon to the self. Chapter 9 discusses how to use theory to understand two important international issues: assessing a country's macroeconomic strategy and evaluating whether a company's international strategy follows the principles of disruptive innovation. This chapter highlights the relationship between disruptive innovation (or the lack of it) at the microeconomic (company) level, and a nation's macroeconomic growth. Finally, chapter

10 returns to the telecommunications industry to show how several emerging technologies have the potential to dramatically reshape the industry.

One important note: Although we make predictions in this book, readers looking for stock picks should look elsewhere. Where we can, we will show indicators and signs that demonstrate an industry turning in a certain direction. But even our very best predictions are contingent on decisions by other firms. We know the forces that act on every firm. With high probability, we can predict how managers will react to those forces. We do believe, however, that managers who understand the forces can account for them and behave differently. An entering firm can do everything right but still get crushed by an incumbent firm that takes the right countervailing options. Companies will make many critical decisions while we breathlessly wait for *Seeing What's Next* to make it to the bookshelves.

That's all right. Our goal is to teach readers how to use the theories of innovation to predict industry change. As the old saying goes, "Give a man a fish and you feed him for a day. Teach him how to fish and you feed him for a lifetime." Our goal is to teach you how to fish.

Readers can use the theories to answer many pressing questions in today's business world. For instance: Why did media conglomerates struggle? Where will convergence succeed? How? Will Linux topple Microsoft? Why or why not? What is the next disruptive wave in retail? Will hyped technologies such as nanotechnology have the effect people think? What path presents the greatest chance of success?

Using theory in a meticulous, rigorous fashion can shine a light where darkness once prevailed. It can thankfully bring an end to an era when hucksters and augurs made their livings selling splendid tales to desperate disciples who needed something, anything, to help guide their decision-making processes. Using theory allows us to see the future more clearly and act more confidently to shape our destiny.

Notes

1. Michael E. Porter, *Competitive Strategy: Techniques for Analyzing Industries and Competitors* (New York: Free Press, 1980).

2. Clayton M. Christensen, *The Innovator's Dilemma: When New Technologies Cause Great Firms to Fail* (Boston: Harvard Business School Press, 1997).

3. Clayton M. Christensen and Michael E. Raynor, *The Innovator's Solution: Using Good Theory to Solve the Dilemmas of Growth* (Boston: Harvard Business School Press, 2003).

4. In more academic parlance, theory is constructed in a cyclical pattern of observation, categorization, prediction, and confirmation. Successful application of this process yields mutually exclusive and comprehensively exhaustive categorization schemes such as: heads or tails; solids, liquids, gasses, or plasma; single, double, triple, or home run. In the business world, these classifications appear as circumstances facing mangers. Theory—a statement of what will result from a certain action in a certain situation—can then help decipher what a company can expect if it takes action *A* during situation *X*.

The process of theory building, however, does not stop with the initial formulation of the theory. The theory must be tested and confirmed. A theory can never be completely proven true; it can only be not proven false. Searching for and identifying anomalies continues to ensure that the theories correctly describe the way the world works and why.

For more information, see Christensen and Raynor, *The Innovator's Solution*, chapter 1; Clayton M. Christensen and Michael E. Raynor, "Why Hard-Nosed Executives Should Care About Management Theory," *Harvard Business Review*, September 2003; Clayton M. Christensen, Paul Carlile, and David Sundahl, "The Process of Theory Building," Unpublished paper. (Paper or electronic copies are available from Professor Christensen's office. E-mail cchristensen@hbs.edu for more information.)

The process of building theory has been articulated by a number of philosophers of science, such as Thomas Kuhn and Karl Popper, and social scientists, such as Karl Weick and Fritz Roethlisberger. See Thomas Kuhn, *The Structure of Scientific Revolutions* (Chicago: University of Chicago Press, 1962); Karl Popper, *The Logic of Scientific Discovery* (New York: Basic Books, 1959); Karl E. Weick, "Theory Construction as Disciplined Imagination," *Academy of Management Review* 14, no. 4 (1989): 516–531; Fritz Roethlisberger, *The Elusive Phenomena* (Boston: Harvard Business School Division of Research, 1977).

5. See Christensen, *The Innovator's Dilemma*; and Christensen and Raynor, *The Innovator's Solution*, chapters 2 and 4. Short summaries of the disruptive innovation theory can be found in: Joseph L. Bower and Clayton M. Christensen, "Disruptive Technologies: Catching the Wave," *Harvard Business Review*, January–February 1995, 43–53; Clayton M. Christensen, Mark W. Johnson, and Darrell K. Rigby, "Foundations for Growth: How to Identify and Build Disruptive New Businesses," *MIT Sloan Management Review* 43, no. 3 (spring 2002).

6. See Christensen, *The Innovator's Dilemma*, 2d ed., chapter 8; Christensen and Raynor, *The Innovator's Solution*, chapter 7; Clayton M. Christensen and Michael Overdorf, "Meeting the Challenge of Disruptive Change," *Harvard Business Review*, March–April 2000, 66–76.

7. See Christensen and Raynor, *The Innovator's Solution*, chapters 5 and 6; Clayton M. Christensen, Michael Raynor, and Matthew Verlinden, "Skate to

Where the Money Will Be," *Harvard Business Review*, November 2001, 72–81; Clayton M. Christensen, Matt Verlinden, and George Westerman, "Disruption, Disintegration, and the Dissipation of Differentiability," *Industrial and Corporate Change* 11, no. 5 (2002): 955–993.

8. IBM began the process of modularizing its architecture with its System 360, which it introduced in 1964.

9. Our repeated use of the word *signal* is important. A signal should not be confused with conclusive evidence. A signal means that there is a chance an industry is in a particular circumstance. Signals are rarely "binary" in the sense that they either exist or do not exist. The goal here is to dramatically increase the odds of getting things right.

10. The history of the telephone draws on a number of very cogent academic works. In alphabetical order: William Paul Barnett, "The Organizational Ecology of the Early American Telephone Industry: A Study of the Technological Cases of Competition and Mutualism" (Ph.D. diss., University of California, Berkeley, 1988); Herbert E. Casson, *The History of the Telephone* (Chicago: A. C. McClurg & Co., 1910), <http://www.the freelibrary.com> (accessed 16 April 2004); Alfred D. Chandler Jr., *The Visible Hand: The Managerial Revolution in American Business* (Cambridge, MA: Belknap Press of Harvard University Press, 1977); Steve Coll, *The Deal of the Century: The Breakup of AT&T* (New York: Atheneum, 1986); Amy Friedlander, *Natural Monopoly and Universal Service: Telephones and Telegraphs in the U.S. Communications Infrastructure 1837–1940* (Reston, VA: Corporation for National Research Initiatives, 1995); Tom Perera, "History, Theory, and Construction of the 'Electric Telegraph' W1tp Telegraph and Scientific Instrument Museums," <http://w1tp.com> (accessed 25 April 2002); George David Smith, *The Anatomy of a Business Strategy: Bell, Western Electric, and the Origins of the American Telephone Industry* (Baltimore: Johns Hopkins University Press, 1985); Neil H. Wasserman, *From Invention to Innovation: Long-Distance Telephone Transmission at the Turn of the Century* (Baltimore: Johns Hopkins University Press, 1985); Kevin G. Wilson, *Deregulating Telecommunications: U.S. and Canadian Telecommunications, 1840–1997* (Lanham, MD: Rowan & Littlefield, 2000).

11. Price adjustment based on Minneapolis Federal Reserve Board, "Consumer Price Index (Estimate) 1800–2000," <http://minneapolisfed.org/research/data/us/calc/hist1800.cfm> (accessed 25 April 2002).

12. Casson, *The History of the Telephone*.

13. Friedlander, *Natural Monopoly and Universal Service*, 42.

14. Chandler, *The Visible Hand*, 197. The merger of three leading telegraph companies in 1866 created Western Union.

15. Ibid.

16. Alexander Graham Bell, "Improvement in Telegraphy," <http://www.iath.virginia.edu/albell/bpat.2.html> (accessed 23 April 2002).

17. Friedlander, *Natural Monopoly and Universal Service*, 28.

18. Historians differ on the specific reason that Western Union decided to exit the local telephony market. Some claim Western Union thought it would lose a patent infringement case filed by Bell. Others claim it was distracted because it was simultaneously fending off a hostile takeover bid from railroad tycoon Jay Gould. But, as discussed, our theories suggest Western Union's decision was a logical response to a disruptive innovation.

19. Malcolm Spicer, "Wireless Deals Fatal Blow to Pay Phones," *Wireless Today*, 6 February 2001. The article notes how the number of pay phone calls dropped by more than 40 percent from 1996 to 2001.

20. Knox Bricken, "Customers Still Holding on to Their 'Trustworthy' Wireline Phone," Yankee Group, 14 May 2002; Eugene Signorini, "The 2000 Mobile User Survey, Part 1: U.S. Wireless Subscriber Preferences and Perceptions at the End of the Second Generation," *Yankee Group Report* 2, no.1 (February 2001); "Bye, Bye Landline Phones," <http://www.cnn.com/2003/TECH/ptech/08/04/cell.only.ap/> (accessed 28 August 2003).

21. Craig McCaw founded McCaw Cellular in the 1980s. His insight was that building a nationwide network would be extremely important. So he moved quickly to acquire licenses in diverse geographic locations at prices that at the time seemed astronomical. However, McCaw's instinct proved correct. AT&T purchased McCaw for an astonishing $11.5 billion in 1994.

22. Wayne Schelle, founder of Cellular One, indicated that businesspeople purchased between 90 to 95 percent of the early cellular systems. Service providers thought businesspeople would be the only ones willing to cover the cost of building out networks capable of covering huge distances and geographic areas. Of course, the only people who really would have cared about a wide-branching network were mobile businesspeople. Schelle later sold Cellular One to Southwestern Bell. See Elisabeth McAllister, "Newest Rage in North Virginia," *Washington Post*, 26 July 1984.

23. This interconnection point is modular in nature and regulated by the FCC. The technology requirements as well as the fees charged for traffic exchange across these interconnection points are well defined, and the resulting interactions are well understood. Wireless technologies' dependence on the wired networks ends at this modular point and a regulated calculation of cost is standard throughout the industry.

PART I

How to Use Theory to Analyze

CHAPTER ONE

THE SIGNALS OF CHANGE

Where Are the Opportunities?

How can we tell that the formula that led to success in the past might not work in the future, or that firms that were successful in the past might not be successful in the future? What specific developments matter the most? What customer groups should we watch to detect such developments? How do contextual factors influence innovation?

The first part of our analytical process shows how to identify the signs that indicate something potentially important is happening. Looking at nonconsumers, undershot customers, and overshot customers and evaluating the nonmarket context for innovation allows us to answer the signals of change question: "Are there signs that someone is capitalizing on opportunities for change?" Figure 1-1 summarizes the topics covered in this chapter.

The first part of using theory to predict industry change involves understanding when to reasonably expect innovation to lead to the emergence of new companies or business models that could be harbingers of industry change.

The core of the signals of change question involves evaluating three customer groups:

FIGURE 1-1

Signals of Change

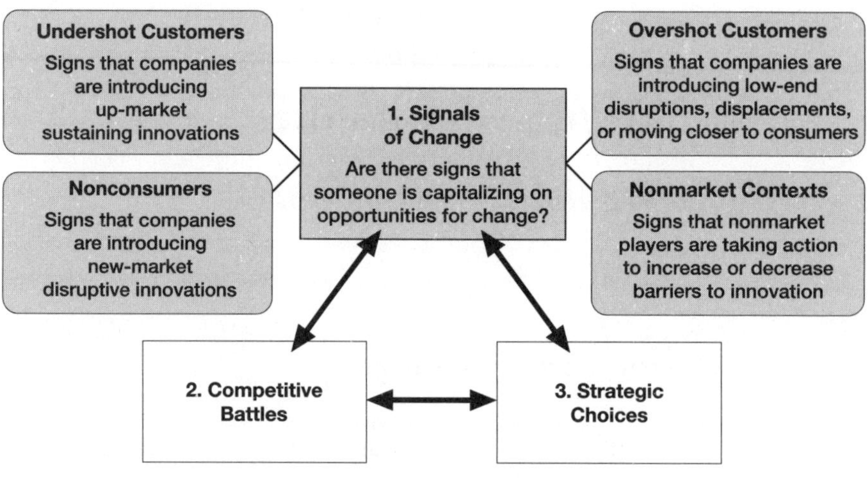

1. Customers not consuming any product or consuming only in inconvenient settings (nonconsumers)
2. Consuming customers who are undershot
3. Consuming customers who are overshot

Each customer group creates unique opportunities. Companies can create new-market disruptive innovations to reach nonconsumers; they can launch up-market sustaining innovations to reach undershot customers; and they can launch low-end disruptive innovations or modular displacements to reach overshot customers. Identifying the industry circumstance is important because it defines what sorts of innovations *will not* flourish. In other words, if circumstances favor up-market sustaining innovations, we should expect firms following a low-end disruptive strategy to struggle.

Most of this chapter focuses on the first and third customer groups. Interestingly, observers often watch developments among a subsegment of the second group known as lead customers—those at the high or performance-demanding end of the market—to assess how a market will change. Sustaining innovations are often deployed there first, and then trickle down into the volume tiers of the market. But for disruptive innovations, the lead customers are in new markets or in the low end of existing markets. Therefore, predicting whether disruptive innovations are taking root, and predicting how they will affect the mainstream of a market in the future requires watching the low end, new markets, and new contexts.

Table 1-1 summarizes these customer groups and shows how to identify each group, the opportunity each group presents, and the signals indicating something or someone is emerging to capitalize on the opportunity.

TABLE 1-1

Overview of Potential Customer Groups

Customer Group	Indentifier	What Could Happen	Signals
Nonconsumers	People who lack the ability, wealth, or access to conveniently and easily accomplish an important job for themselves; they typically hire someone to do the job for them or cobble together a less-than-adequate solution	New-market disruptive innovation	• Product/service that helps people do more conveniently what they are already trying to get done • Explosive *rate* of growth in new market or new context of use
Undershot customers	Consumers who consume a product but are frustrated with its limitations; they display willingness to pay more for enhancements along dimensions most important to them	Sustaining up-market innovation (radical and incremental)	• New, improved products and services introduced to existing customers • Integrated companies thrive; specialist companies struggle
Overshot customers	Customers who stop paying for further improvements in performance that historically had merited attractive price premiums	Low-end disruptive innovations	• New business model emerges to serve least-demanding customers
		Displacing innovation	• Emergence of specialist company targeting mainstream customers
		Downward migration of required skills	• Emergence of rules and standards—widely propagated statements of what causes what • Migration of provider closer to end customer

Nonconsumers and Opportunities for New-Market Disruptive Growth

The first group of customers to look for, interestingly, is people who are *not* consuming. Nonconsumers exist when characteristics of existing products limit consumption to people who have significant financial resources or specialized skills or training. Nonconsumers have a job they need to get done. But they are left on the sidelines, unable to achieve the outcome they desire satisfactorily. No existing market offering is designed to serve them. What can they do? They either pay professionals to provide the service for them, or do the best they can to cobble together a solution from existing products and services.

In the 1870s, most people were trying to communicate over long distances, as evidenced by the letters they wrote. Few of them had the skill to communicate by telegraph—and doing so was expensive and inconvenient. It required traveling to a central facility and working with an expert operator trained in Morse code. If the expert happened not to be there (and you could not call ahead to check), you were out of luck. The telephone competed against nonconsumption because it enabled individuals to communicate over longer distances without an expert's assistance. The fact that its signal carried only a few miles didn't matter, because the telephone initially didn't compete against the telegraph.

Nonconsumers are ubiquitous. They exist in every market. Even people consuming a product can be nonconsumers. How? A person might not be consuming a product in a particular context or environment. For example, in the 1980s, almost every U.S. resident used landline telephone service. Whether they were at home or in the office, they could simply pick up the nearest receiver and hear a familiar dial tone. However, most people were not consuming telephony services in mobile contexts. Public pay phones were typically inconvenient, if they worked at all. In the mobile environment, most people were nonconsumers.

Because identifying nonconsumption is easy, the important issue is whether a company is doing anything about it. Companies can reach nonconsumers with new-market disruptive innovations such as the telephone or wireless telephony. Successful new-market disruptive innovations follow two patterns:

1. They introduce a relatively simple, affordable product or service that increases access and ability by making it easier for cus-

tomers who historically lacked the money or skills to get important jobs done.

2. They help customers do more easily and effectively what they were already trying to get done instead of forcing them to change behavior or adopt new priorities.

The first element of the pattern points to the importance of competing against nonconsumption rather than competing against consumption. New-market disruptive innovations lack the raw functionality of existing products but bring new benefits such as convenience, customization, or lower prices. The attribute bundle means the product will only find success if it takes root among new customers or in a new context of use. Demanding customers who are already consuming a potentially competing project will reject the innovation because of its performance limitations. Competing against nonconsumption entails clearing a lower acceptance hurdle.

For example, one natural entry strategy for emerging telephony companies in the 1870s would have been to compete against consumption and target telegraphy users. But the original technology was too limited to be of much use to demanding communications customers. Sophisticated telegraph users would have instantly rejected the telephone because it could not transmit information over long distances. However, in a different context—local communications—there was no real competition to beat; to talk to someone beyond shouting distance, you either had to walk or ride a horse to get to that person or make do without communicating at all. The telephone just had to be better than these options in order to delight the customer.

Following the second element of the pattern requires that companies make it easier for consumers to accomplish important unfulfilled (and often overlooked) outcomes—jobs people need to get done but can't.[1] The telephone made it easier for people to communicate over short distances using the familiar nature of speech. Users had to pick up a phone and ask an operator to connect the call. Later, when wireless phones were introduced, they also clearly fit with customers' existing behavior patterns. Customers turned on a familiar-looking device, dialed a number on a standard keypad, and spoke into a receiver. Even the basic features and pricing model based on minutes of use mimicked the familiar landline alternative. The only major difference was the absence of a dial tone. Wireless phones made it easier for people to do jobs they historically prioritized but could not adequately satisfy, such as "make my commute more productive" or "protect me in case of emergency."

New-market disruptive innovations have the greatest potential for long-term industry change. However, they are the hardest innovations to identify. What are the signals that a company is creating new-market disruptive growth? One clear signal is a high and increasing rate of growth in a new, emerging market. The key is looking beyond the sheer size of a new market to the growth rate and increases to the growth rate. Identifying new markets with increasing rates of growth can allow you to spot important developments while they are still quite small.[2] Both the telephone and wireless phone created strong growth in new communications contexts. Another signal is action in targeted customer segments—for example, teenagers, college students, hackers, small business owners, or people in developing countries. Each of these groups typically puts up with performance imperfections when a new product or service makes it easier to do something they could not do in the past.

How can you identify whether nonconsumers exist? One way is to map the product or service delivery chain. New-market disruptive innovations tend to take a link out of this chain—allowing people to do for themselves what previously required expertise. The right kind of market research that seeks to identify unfulfilled jobs can also be quite useful.[3]

One note about price: New-market disruptions are always relatively low-priced. They are not always cheap on an absolute scale, however. The first mobile phones, personal computers, cameras, and so on—all were expensive. However, they all were significantly more affordable than available technological solutions. For example, the only real alternative to a mobile phone in the late 1970s was to provide CB radios to everyone you hoped to contact. That solution was prohibitively expensive, highly inconvenient, and difficult to implement. The expensive nature of some new products limits consumption to people who desperately need to get a job done. Subsequent improvements typically create production efficiencies that enable price reductions that make the disruptive product or service available to wider customer groups.

Undershot Consumers and Opportunities for Up-Market Sustaining Innovations

After identifying whether companies are establishing new ways to reach nonconsumers, the next step is to evaluate *current* customers. Every market consists of multiple customer tiers. At the high end of a market are demanding customers who have very tough problems to solve. At the low end of the market are less demanding customers who have relatively fewer or less complex requirements to satisfy. Our theories separate existing customers into two classes: undershot customers, for whom existing products are not good enough, and overshot customers, for whom existing products are more than good enough.

A natural question results from this distinction: Not good enough to do what? Broadly speaking, we mean not good enough along whichever dimension of performance customers in a given tier of the market care about the most. This dimension is an industry's *basis of competition*.[4]

In a product's early days, customers tend to value the dimensions of its performance that determine what the product allows them to do (functionality) and how consistently it gets the job done (reliability). Companies that come closest to meeting customers' requirements or add some further level of functionality or reliability can earn above-average profits.

Numerous signals indicate undershot customers. Consumers expressing frustration, product reviews highlighting industry-wide limitations, and articles littered with language such as "if only" all point to undershot customers. Undershot customers frequently turn to aftermarket consultants who must resort to creative solutions to rig the product or service to provide the extra performance customers need. The most telling signs of undershot customers are markets in which customers are consistently willing to pay premium prices to trade up to new, higher-performing products. Other signals can be the prosperity of integrated companies that provide full system solutions and the struggles of specialist companies that do not have the requisite capabilities to solve complex, interdependent problems.

The telephone's initial limited range and reliability undershot its early customers. Customers eagerly welcomed and willingly paid for improvements in reliability and range.

Undershot customers create opportunities for existing firms to profitably introduce up-market sustaining innovations. These sustaining innovations make good products better. In this context, companies can safely offer improved services at more attractive prices to their most important customers.

Up-market sustaining innovations fall on a continuum between radical and incremental improvements.[5] Radical sustaining innovations are at the complex end of the continuum. These "great leaps forward" tend to be complicated, interdependent, and expensive. Two classic examples of radical sustaining innovations are the systemwide upgrade of the telecommunications network from analog to digital technology and the move from black-and-white to color televisions.[6] Incremental sustaining innovations tend to influence an industry less dramatically. Early innovations such as the switchboard and loading coils improved the telephone's performance, but we would classify them as incremental innovations because they did not entail significant system reconfiguration.[7] Many of the telecommunications features that we now take for granted—such as broken dial tones when we receive a voicemail message and caller identification—can be classified as incremental sustaining innovations.[8]

Integrated companies tend to be good at both forms of up-market sustaining innovations. Integration is an absolute necessity for radical sustaining innovations. Integrated companies can master the myriad interdependencies involved in wrestling with compatibility, interoperability, and legacy issues. Specialist companies just don't control enough pieces of the puzzle to effectively commercialize radical sustaining innovations.

For example, vertical and horizontal integration helped AT&T beat not only Western Union but also the hordes of other local telephony companies that had come to populate the industry in the late 1800s. In 1881, the Bell Company purchased leading telecommunications equipment provider Western Electric. In 1885, it reorganized under the AT&T mantle. Under the stewardship of Theodore Vail, AT&T bought the Bell licensees out in 1899, and it subsequently snatched up all but a few independent local telephone companies.

Although some might view Vail's actions as a greedy grab for monopolistic power, there actually were technological reasons why an integrated giant *had* to emerge at that time in telephony. Bell's licensing model led to the creation of thousands of local operators and numerous equipment suppliers. Between 1894 and 1904, more than six thousand independent telephone companies went into business in the United States. This hodgepodge of suppliers and providers created overwhelming management challenges. Coordination was difficult, network monitoring was next to impossible, operators experienced diseconomies of scale, and service quality suffered. Only integration could solve these sorts of problems. Integration gave the Bell Company, and later AT&T, crucial advantages over nonintegrated competitors. For example, owning its equipment supplier, Western Electric, allowed the Bell Company to ensure the interoperability of every piece of equipment on its network, improving the predictability of interactions between equipment and increasing the network's reliability.

Centralization coupled with AT&T's size brought the fear of monopolization. Nonetheless, the fully integrated system was critical to developing solutions for many of the problems that plagued the loosely coupled Bell network. AT&T was able to provide substantially better service than independent phone companies working with independent equipment suppliers. Hence, Theodore Vail's efforts to integrate AT&T made sense as a way to manage the complex interdependencies within the system to ensure a reliable phone connection for every American.[9] Vail responded to the same mandate that led U.S. Steel, RCA, IBM, Intel, and Microsoft to become dominant, integrated companies at a similar stage in their respective industries' histories.[10]

The sustaining innovations that companies introduce to reach undershot customers are the means by which companies exploit their growth potential after they establish their initial foothold. They are not a focus of this book, however. Many of the classic analytical techniques people use to predict how new technologies will affect the competitive structure of industries are valuable tools to understand the impact of sustaining innovations because they take place in existing, measurable markets and offer improvements along established performance metrics.[11]

Overshot Consumers and Opportunities for Low-End Disruption, Shifting Profits, and Emergence of Rules

The final group of existing customers is overshot customers. As companies introduce up-market sustaining innovations and improve their products and services, they eventually overshoot the performance that some of their customers can use. One bedrock finding from our research is that companies innovate faster than customers' lives change. In other words, what people are looking to get done remains remarkably consistent, but products always improve. Thus, products eventually become too good. Overshooting is the driver behind commoditization—the process that results in companies being unable to profitably differentiate their products and services. If overshooting never occurred, products would never mature. Customers would always be willing to pay higher prices for better products.

Overshooting redefines the type of innovation that can create opportunities for growth because companies can alter an industry's basis of competition. It opens the door for several important forms of industry change and leads to changes in how companies ought to organize.

How can you tell if customers are overshot? Overshot customers pay decreasing premiums for improvements they used to value. In economic parlance, they derive diminishing marginal benefits from product enhancements. Companies add extra features that never get used. People begin to complain about things they ignored in the past. "This product is too complicated," they say. "And it costs too much."

Nature abhors a vacuum, especially when avenues for profitable growth remain. You can always predict that competitors will be motivated to find new ways to earn above-average profits. After functionality and reliability have become good enough, the next dimensions along which firms can compete relate to *ease of use*—how flexible and easy it is to use a product (convenience), how squarely a product lines up with individual customers' idiosyncratic jobs (customization), and how much it costs to use a product (price). Note that price appears last on this list even though the emergence of price-based competition is the signal most often associated with overshooting. Remember, price-based competition means that companies can no longer earn premium prices for improving a *particular* dimension. It is only after companies fulfill *all* customer needs that price becomes the *only* thing that matters. In other stages, customers reward companies with premium prices for providing superior functionality, reliability, convenience, or customization.

It is important to remember that all customers in a market do not become overshot at once. The condition starts at the bottom of the market and then creeps upward.

Overshooting and the resultant changes in the basis of competition open the door for three different forms of industry change:

1. Low-end disruptions taking root among the most overshot customers
2. Specialists entering and displacing integrated players
3. Standards or rules developing that allow different types of providers to create products and services good enough to meet the minimum requirements of customer segments

We will discuss each of these changes and show how they all require changes in an industry's value chain.

Targeting the Lowest Tier: Low-End Disruptive Innovations

Although they don't create new-growth *markets*, innovators in overshot market tiers can create new-growth *companies* by using low-end disruptive innovations to establish a beachhead among the incumbent's least demanding customers. These customers are the most overshot. They are likely using the incumbent's product or service because it is the only available alternative. But they are unsatisfied. They are paying for functionality and features that just aren't important to them. They are the most likely to desert the incumbent for a company offering lower prices or more convenience.

For example, MCI created low-end disruptive growth by targeting price-sensitive customers who did not need all of the functionality offered by AT&T.[12] MCI's Execunet service, introduced in the 1970s, allowed business users to pay lower prices if they were willing to dial a twenty-two-digit code to connect to MCI's network. MCI built its own long-distance network and relied on AT&T's local lines for call origination and completion. Although AT&T predictably protested the requirement that it allow its competitor to use its local network, MCI could plug its equipment into AT&T's switches relatively easily. Regulation granted MCI permission to enter at this point of modularity.

Although MCI did not (and could not) offer all of the advanced features that AT&T's customers enjoyed, it offered much lower prices than AT&T. Regulation prevented AT&T from dropping its prices without appealing to the local regulatory commission.[13] Exploiting this price discrepancy, MCI began to build a large customer base by reaching price-sensitive business customers with its low-price offering.[14] MCI's customers tolerated lower functionality in exchange for lower prices.

A signal that a company is launching a low-end disruptive innovation is the creation of a business model that makes money in a different way than established companies—for instance, lower prices but higher asset turnover, a different mix of sales and post-sales support revenue, and so on.

Entering as a Specialist: Displacements

Specialist providers can introduce a *displacing* innovation and take part of a market from an incumbent. Displacements are a distinct class of innovation. Unlike up-market sustaining innovations, displacements take place at a point of *modularity*. Unlike low-end disruptions that first target the least demanding customers, displacements first target the mainstream market. Displacements do not necessarily involve low-cost business models or products with performance limitations. Specialists who focus on one particular piece of a product or service tend to introduce displacements.

Consider the emergence of competition in the customer premise equipment (CPE) market. In the 1950s, Western Electric (AT&T's equipment arm) controlled the residential CPE market. As implied by the name, CPE is anything a customer plugs into the network and uses, such as a standard telephone. Western Electric produced a phone that was indestructible and available in any color you wanted—as long as you wanted black or beige. Efforts to sell non–Western Electric CPEs were restricted, because nonsanctioned equipment could threaten the telephone network's reliability.[15] In simple terms, plugging something else in might unexpectedly affect the entire network, so plugging in only those technologies designed and made by AT&T was the best alternative.

In the late 1950s, entrepreneur Tom Carter introduced the Carterfone.[16] Resembling a walkie-talkie, the Carterfone transmitted a voice signal from a remote device to a speaker placed next to the phone. It let farmers use their telephones when they were in the field. Because the Carterfone had a direct electrical connection to AT&T's network, AT&T cut off service to any subscriber it found using the device.

After an FCC decision in 1968 sanctioning the use of the Carterfone and a protracted legal battle that eventually reached the Supreme Court, the government decided that consumers could attach devices to the edge of the network that met a set of specifications and that did not cause any unanticipated performance problems elsewhere in the system.

The court decision opened the floodgates. New specialist companies offering a wide variety of devices entered the CPE market, stealing substantial share from Western Electric. Competition eventually led to the creation of innovative types of CPE that created booming new product segments, such as the fax machine, the modem, and the private-branch exchange (PBX), which manages and directs calls in most modern businesses. Specialist companies arose and captured these new markets.[17]

The financial services industry has also seen a number of displacements. Historically, the same financial institution would originate, service, and source a loan (such as a mortgage, auto loan, or consumer credit card). The past two decades have seen the emergence of specialist providers. Asset securitization and credit scoring (discussed below) have fueled this displacement.

When looking for displacements, look for functionality that overshoots customers *and* for modular interfaces. Both are important: Remember, AT&T and Western Electric didn't immediately lose *all* of their business when competition emerged in the CPE market. Specialist companies had no chance of competing against AT&T's equipment at the core of the telephone system. AT&T's integration still served as a source of unparalleled competitive advantage. Only AT&T could develop and master the intricate interdependencies involved in the electronic switches that constituted its core operations. As a general rule, specialists cannot win unless their products can interface with the larger product system at points of well-defined modularity. For example, the bevy of specialist firms spawned by the Telecommunications Act of 1996 (Telecommunications Reform Act or TRA)—particularly the so-called competitive local exchange carriers (CLECs)—struggled. The CLECs found it surprisingly difficult to plug into the local telephony provider's networks. There was no clean interface allowing easy entrance. (We will discuss the act and the CLECs in detail in chapter 4.)

What indicates a modular, or clean, interface? Chapters 5 and 6 of *The Innovator's Solution* noted three tests that determine modular interfaces:

1. Managers must be able to specify which aspects of the interface are important and which are not.
2. Managers must be able to measure or verify that the parameters and exchanges that comprise this interface are correct and are what they need.
3. The interactions across the interface must be well understood and predictable. As long as there are unpredictable interactions between components, efforts to move toward modularity have potentially disastrous consequences.

One final note about displacements: Because displacements result in specialization, they can facilitate low-end disruption. How? A new company can emerge and piece together value chain components in new ways to offer new benefits. For example, Dell took advantage of the modular nature of the personal computer industry when developing its low-end disruptive business model. In an integrated world, Dell would not have been able to enter in this way.

Development of Rules Allows Migration of Producer Toward End-User

The final type of change that can occur when customers are overshot is the emergence of rules allowing producers closer to the end consumer with less skill to produce good-enough products. Great science does indeed simplify. It opens the door for both new-market and low-end disruptive innovations as new industry players have a way to provide products in a new context or piece together a business model that offers a good-enough product at radically lower cost.

The pioneering companies in almost every industry have to solve their technological and product development problems in an unstructured, experimental manner. The companies do not know what causes what and why. A company in this state places a high premium on expert knowledge or scientific skill.

As companies gain experience solving problems, they begin to see patterns of cause and effect. Ultimately, the system becomes well-enough understood that rules can emerge to guide development efforts. Eventually, rules become so widely accepted that people begin to see them as standards. When a product is more than good enough, these rules or standards allow people with less expertise to follow the rules and produce a good-enough product—something that earlier would have required deep expertise.

The propagation of widely accepted standards by which components of a system are to interface is one signal that this change has occurred. Another signal is when a company places less emphasis on deep theoretical knowledge about a subject matter in its hiring practice.

Interestingly, companies marching along sustaining improvement trajectories often unwittingly develop the rules that put their competition into business. Often, a company will develop design rules to speed its development process. Rules allow the company to partition the problem, enabling separate groups to work independently on individual components or subsystems, simply plugging them into each other at the end. However, defining these interfaces enables nonintegrated companies to begin producing subsystems, and companies with less technical depth to become assemblers of modular products.

From the perspective of an incumbent, this could appear as new-market disruptive growth (because of the entrance of a provider previously locked out of a market), but from a consumer perspective, it could appear to be low-end disruptive growth (because they get a cheaper product). This is one reason why it is useful to think of low-end and new-market disruptive innovations as ends along a continuum.[18] The emergence of rules that allow a shift closer to the consumer enables the emergence of companies that sit squarely in the middle of this continuum. In other words, new companies that take advantage of rules have elements of both low-end and new-market disruptions.

The development of rules or standards that defined the interface between stages of value added facilitated the banking industry's vertical disintegration. When banks first had to assess customers' creditworthiness, they had no choice but to rely on a loan officer's expert judgment. The loan officer looked over all of the customer's financial records and interviewed the customer. Based on this, the loan officer developed an intuitive feel for whether the bank could extend credit to the customer.

Over time, people began to recognize patterns. Four variables really made the difference between good risks and bad risks: length of time at current residence, length of tenure with current employer, annual income, and past performance in bill payment. In 1956, Fair, Isaac created a standard, predictive risk-assessment tool. It dramatically simplified the process of judging creditworthiness with a statistical methodology that plugged variables from an applicant's credit history into an algorithmic formula that produced a score. Credit scoring's robust, scientifically based, quick assessment enabled a broader population of less-skilled people to make lending decisions.

The first adopters of credit scoring were department stores and oil companies that issued their own credit cards. Credit scoring quickened the process of extending small loans to consumers. Further refinements to the methodology led to its use for general credit cards, auto loans, home mortgage loans, and, most recently, small business loans. The development of rules allowed loan decisions to move closer to the ultimate consumer, enabling different types of producers—such as specialist credit card providers (for example, MBNA) and other specialist providers without expert risk-assessment knowledge or pools of loan officers—to offer more than good-enough service and increasing customization and convenience.

One note: We observe a tendency for outsiders to applaud any time an industry gravitates toward standards. Standards are certainly good when they allow companies to quickly develop good-enough products and services. However, standards typically involve compromises and trade-offs. They pull back from the edge of what is possible in the interest of promoting speed and flexibility. When there are undershot customers, standards inhibit the ability of companies to produce the best products possible. In other words, standards are not always a good thing.

Changing the Basis of Competition Requires Changes in Integration

Companies leading any of the changes discussed in this section organize in much different ways than established industry leaders. The value chain evolution (VCE) theory explains why. Remember the theory's golden rule: Organizations need to integrate across the interfaces in the value chain that drive the performance of what is not good enough. When the functionality and reliability of products overshoot customer needs, then convenience, customization, and low prices become what are not good enough. Just as delivering maximum functionality and reliability requires a particular type of integration and organization, delivering these new benefits requires integration across different stages in the value chain. In other words, the necessary locus of integration shifts when the hard problem shifts.

When functionality and reliability are inadequate, companies that try to maximize performance typically integrate the critical elements of design as well as production of a product or service. IBM controlled every aspect of the original mainframe. AT&T controlled the end-to-end telecommunications network.

When companies solve the hard problems related to performance, this form of integration is no longer necessary. Specialist companies can provide good-enough pieces of a product or service at defined points of modularity. Profitability flows away from firms that assemble modular products or services (they become "thin" integrators) to firms that produce critical subsystems, and to firms that integrate at points that determine speed- and convenience-defining improvements.

For example, Dell introduced new means of convenience and customization to the personal computer market. To drive performance improvements along those metrics, Dell integrated across the supply chain and the customer relationship through its direct ordering system, because that is where leverage over the ability to customize rapidly and conveniently is determined. Wal-Mart integrated similar pieces of its value chain to deliver its low-cost business model. Specialist CPE providers captured value in the 1960s and 1970s. All of these companies organized in a very different manner from the prior market leaders.

Conservation of Integration

The shift of one portion of a value chain from integration to modularity affects the rest of the value chain. Why? Integration is like energy: It doesn't go away. This *conservation of integration* (also known as "the law of conservation of attractive profits") helps explain how value chains reconfigure to support an industry's basis of competition. We will describe it briefly here and use it in great depth to look into the future of the semiconductor industry in chapter 7.

Conservation of integration holds that, when an interdependent system architecture is necessary to optimize performance at a stage of value added that is not good enough, the architecture of the product or service at the adjacent stage of value added must be modular and conformable in order to optimize the performance of what is not good enough.[19] In simple terms, modular stuff must surround integrated stuff to optimize the integrated stuff.

Consider the difference between Microsoft Windows and the Linux operating system. Windows is a highly integrated, interdependent operating system. To optimize the operating system, applications developers must conform their products to meet Microsoft's interface requirements. Efforts to try to modify Windows to improve individual applications would be disastrous; any individual change would have literally thousands of unanticipated consequences and operating system problems. Linux works the other way, because its goal is to enable optimized applications. The Linux operating system itself is modular. As long as you follow the rules, you can modify it to optimize the performance of an application.

We have also called the law of conservation of integration the "law of conservation of attractive profits," because companies make attractive money when they solve the hardest problems. The hardest technical problems mandate solutions that are tightly coupled integrated systems. When modularity and commoditization cause attractive profits to disappear at one stage in the value chain, conservation of integration means the opportunity to earn attractive profits with proprietary products will emerge at an adjacent stage.

Conservation of integration explains why it is not accurate to characterize an industry as *integrated* or *disintegrated*. There are numerous types of integration in any given industry. Specialists *are* integrated, just in different ways. Whereas integrated firms might span an entire industry's value chain, specialist firms might be integrated to produce a single component that itself is not good enough. Or they could be integrated across whatever interfaces drive customization and convenience, which could be the point of customer interaction or the interface with suppliers.

The Important Nonmarket Context for Innovation

Nonconsumers, undershot customers, and overshot customers all present opportunities for new companies or new business models to emerge. Before taking the next step and showing how to use our theories to evaluate emerging head-to-head battles, we must discuss one more topic. Innovation always takes place within a context. Nonmarket forces, notably the government and its regulatory agencies, can play a very important role in shaping that context. Thus, this chapter's final section briefly shows how to understand the nonmarket forces that influence opportunities for change. Because we have not written extensively about this specific topic before, chapter 4 develops the motivation/ability framework in more depth, illustrating how to assess whether actions by nonmarket players are likely to have their intended impact. This chapter provides a brief overview of that framework because it forms a critical component of the signals of change question.

Some of our most important industries, such as telecommunications, health care, financial services, and education, feature involvement by the government and its regulatory bodies. Comprehensively looking into the future of these industries requires a way to analyze how nonmarket players affect innovation. The simple answer that "less is more"—less regulation means good things and more regulation means bad things—just doesn't suffice.

Two factors exist in abundance in environments where innovation thrives: motivation, defined as market incentives to innovate; and ability, defined as the capability to obtain resources, craft them into products and services, and offer those products and services to customers. There are ostensibly free markets where motivation and ability exist in abundance. But a spectrum of nonmarket factors—industry standards, unions, cultural norms, the state of technological development, a country's intellectual property infrastructure, and most important, government regulation—affect the motivation and ability to innovate.

Simply stated, the motivation/ability framework suggests that innovation flourishes when companies have both the motivation and ability to innovate. Market contexts that are deficient in motivation or restrict ability are stifling to innovation. Companies that bring nascent innovations to unfavorable market environments end up either searching for environments that are more favorable or abandoning their efforts.

The government's power to affect innovation lies in its policymaking (including subsidies and incentives) and regulatory authority. The government or other nonmarket players can affect either the motivation or ability of industry participants. In doing so, they can change an industry's context, making it more or less conducive to innovation.

Using the motivation/ability framework involves following these three steps:

1. Map existing motivation and ability to see if the context is favorable to the different types of innovation discussed above. If it is not, identify the primary barriers to innovation.

2. Determine if any nonmarket player is taking action to influence motivation and ability.
3. Determine if the action addresses the primary barriers to innovation. If it does, expect the action to promote innovation.

When the government or another nonmarket player takes actions that address critical deficiencies in motivation or ability, the market for innovation improves. For example, in the 1960s and 1970s the primary barrier to innovation and competition in the CPE and long-distance markets was access to customers. Firms were motivated to enter both markets. They had developed innovations they thought customers would value. But existing regulations prevented them from reaching customers. Government efforts to promote ability increased the pace of innovation.

When the government takes actions that fail to address the root cause of the deficiency or actually increase the deficiency, the market for innovation tends not to improve. For example, chapter 4 will explain in depth how government efforts to create competition in the local telephony market struggled because the government could not address some of the real barriers to innovation.

The first part of using the theories of innovation to analyze industry change requires looking for signals of change—signs that someone is capitalizing on opportunities for change.

Companies can create new-market disruptive innovations to reach nonconsumers. They can introduce up-market sustaining innovations to reach undershot customers. They can introduce low-end disruptive innovations or modular displacements to reach overshot customers. The emergence of rules facilitates these changes. Nonmarket actions that increase motivation and ability can encourage these innovations. All of these developments are signals of change that have the potential to radically reshape an industry's structure. Each occurs in a particular circumstance, meaning we can predict that a company attempting an innovation in the "wrong" circumstance will fail.

Key questions that help to unearth such developments include:

- *What jobs are customers in the industry trying to get done? Are customers not served, undershot, or overshot by cur-*

rent offerings? Along which dimension do firms compete for customers?

- *What improvements garnered premium prices in the past?*
- *Do integrated or specialized business models currently prevail? Are interfaces specifiable, verifiable, and predictable? If so, where is modularity occurring?*
- *Where are new business models emerging? Is there growth in fringe markets?*
- *What role does the government or its regulatory bodies play in enhancing or inhibiting innovation?*

Companies taking advantage of these circumstances are poised to grow and change an industry. That growth usually leads to an entrant encroaching on an incumbent's market. The next chapter shows how to use the theories of innovation to evaluate competitive battles.

Notes

1. Chapter 3 of Clayton M. Christensen and Michael E. Raynor, *The Innovator's Solution: Using Good Theory to Solve the Dilemmas of Growth* (Boston: Harvard Business School Press, 2003) discusses this concept in much greater detail. It explains how customers hire products to do jobs that occur in their lives. We discuss the notion of jobs to be done in more detail in chapter 5 and in the appendix.

2. One way to analyze the rate at which an innovation is growing is to plot a substitution curve. On the *Y*-axis, plot the market share of the new innovation divided by the market share of existing solutions. A figure of 1.0 means the new innovation has 50 percent of the market. A figure of 0.1 means the new innovation has about 9 percent of the market. The *Y*-axis should be on a logarithmic scale. The *X*-axis should plot increments of time, usually in years, on an arithmetic scale. Typically, a straight line will connect the data points. This allows you to assess the rate at which a new technology will take a market with very limited data. For an example of what these curves are and how to construct them, see Clayton M. Christensen, *The Innovator's Dilemma: When New Technologies Cause Great Firms to Fail*, 2d ed. (New York: HarperBusiness, 2000), 184–186.

3. A useful reference on this topic is Dorothy Leonard-Barton, *Wellsprings of Knowledge: Building and Sustaining the Sources of Innovation* (Boston: Harvard Business School Press, 1995).

4. This notion was originally based on a buying hierarchy devised by Windermere Associates (see Christensen, *The Innovator's Dilemma*, 2d ed., 197–198). An unpublished working paper discusses these concepts in greater detail: Clayton M. Christensen, Scott D. Anthony, and David Sundahl, "Drivers of Change in the Basis of Competition," 2001.

5. There are many ways to categorize sustaining innovations. Kim Clark and Rebecca Henderson developed one particularly helpful classification scheme. Their scheme involved a two-by-two matrix, with the technological concept (new versus same) on the vertical axis and the architectural configuration (new versus same) on the horizontal axis. They termed a new technological concept with a new architectural configuration a *radical* innovation, a new technological concept with the same architectural configuration a *modular* innovation, the same technological configuration with a new architectural configuration an *architectural* innovation, and the same technological configuration with the same architectural configuration an *incremental* innovation. They found that incumbents tended to stumble when faced with architectural change. This was because the incumbents' product development process came to mirror their product architecture. Developing a new architecture required new patterns of coordination and interaction. See Rebecca M. Henderson and Kim B. Clark, "Architectural Innovation: The Reconfiguration of Existing Systems and the Failure of Established Firms," *Administrative Science Quarterly* 35 (1990): 9–30.

Another way to slice sustaining innovations is to separate sustaining innovations that improve a product's performance from sustaining innovations that improve operational effectiveness and boost overall profit margins. For our purposes, we have simplified this into the radical and incremental sustaining innovations discussed in this chapter.

6. Using digital bits instead of analog sound waves allowed for higher transmission speeds, clearer voice quality, and fewer errors. The upgrade provided significantly improved services to customers and sharply lower operating costs to service providers. It was a very difficult and expensive task. Telephone companies had to pull out old, fully depreciated equipment like electromechanical switches throughout the network and install expensive new solid-state devices. The upgrade required the deployment and integration of hundreds of different components and subsystems. Engineers spent countless hours devising novel solutions to improve the performance of what most considered the best-performing system in the world. A nonintegrated company could not have done this.

At the time of the divestiture of AT&T in 1984, capital markets rationally placed different valuations on the operating companies based on their digital deployment. Companies that had upgraded a significant portion of their networks would not only have lower capital expenditures in the future; they would also have higher gross profit margins.

Similarly, the move from black-and-white to color television was a laborious process that required an end-to-end, systemwide overhaul. Broadcasters wouldn't buy equipment to broadcast in color unless customers had sets that would receive in color. Because RCA owned NBC, RCA was uniquely positioned to break the chicken-and-egg dilemma and introduce color televisions.

7. Switching, the ability to route calls through the network, made it even easier to call anyone on the network without requiring a dedicated connection. Calls enter a centralized location and get switched to another line on the network. Loading coils, an induction device, helped to overcome the range problem. The

loading coil's perfection in 1899 effectively doubled the distance a signal could travel without any degradation. By increasing the distance that a call could travel and making it easier to add new users to the network, the Bell companies could attract additional customers, earn more profits, and so on.

8. These incremental sustaining innovations came about as service providers utilized so-called out-of-band signals to provide extra functionality to their customers. Interestingly, they rest on an innovation that could be classified as a radical sustaining innovation—the implementation of the Signaling System 7 (SS7) protocol. SS7 was very technologically challenging to implement, but granted the telephone network much greater functionality.

9. Many historians suggest that a monopoly had to emerge because the telephone system is, in fact, a *natural monopoly*. A natural monopoly is commonly defined as a market situation wherein a single producer is able to obtain an average cost per unit of production sufficiently low to make it inefficient for any other producer to supply that product or service. Natural monopolies typically occur in markets where there is a large fixed capital component coupled with steadily declining average marginal production costs, allowing one supplier to attain unmatchable production advantages.

However, the fact that independent phone companies outnumbered Bell-controlled companies in 1903 raises some doubts about whether the telephone system is indeed a natural monopoly. In fact, dual service was available in about 60 percent of American cities with populations larger than five thousand. It was common to observe multiple sets of telephone wires strung between buildings in dense urban areas.

For more information on natural monopolies and their implications, see Stephen G. Breyer, *Regulation and its Reform* (Cambridge, MA: Harvard University Press, 1982); Thomas J. Dusterberg and Kenneth Gordon, *Competition and Deregulation in Telecommunications: The Case for New Paradigm* (Indianapolis, IN: The Hudson Institute, 1997); Alfred E. Kahn, *The Economics of Regulation: Principles and Institutions* (New York: Wiley, 1970); Richard A. Posner, *Natural Monopoly and its Regulation* (Washington, DC: Cato Institute, 1999); William W. Sharkey, *The Theory of Natural Monopoly* (New York: Cambridge University Press, 1982); John T. Wenders, *The Economics of Telecommunications: Theory and Policy* (Cambridge, MA: Ballinger, 1987).

10. Alfred Chandler was the first to observe this in his seminal work *The Visible Hand*. Chandler observed that for many industries, it was not until a company snatched all of the power in an industry that the industry took off. Our contribution to Chandler's observation is a causal mechanism about why this has to occur, because only technological integration can solve interdependencies and improve what is not yet good enough. See Alfred D. Chandler Jr., *The Visible Hand: The Managerial Revolution in American Business* (Cambridge, MA: Belknap Press of Harvard University Press, 1977).

11. Some helpful sources include Steven C. Wheelwright and Kim B. Clark, *Revolutionizing Product Development: Quantum Leaps in Speed, Efficiency and Quality* (New York: Free Press, 1992); Stefan Thomke, *Experimentation Matters:*

Unlocking the Potential of New Technologies for Innovation (Boston: Harvard Business School Press, 2003); Eric von Hippel, *The Sources of Innovation* (New York: Oxford University Press, 1988).

12. The erosion of AT&T's long-distance monopoly began innocently enough with an FCC decision in 1959 that became known as the "Above 890" decision. The decision allowed firms to use microwave frequencies above 890 megahertz for their private communications. In 1963, MCI petitioned the FCC to allow it to build a shared private network between Chicago and Saint Louis using microwave technology originally developed for military purposes. The private network would only be used for firm-to-firm traffic. It was shared because multiple firms would use the same network. The FCC decided in the late 1960s to allow a little competition at the fringes of the long-distance market, never intending its decision to lead to the breakup of AT&T's monopoly. AT&T challenged MCI in the courts but, based on the FCC's original ruling, a judge decided that MCI could compete with AT&T, opening the door to true competition in long-distance services. See Philip L. Cantelon, *The History of MCI 1968–1988, The Early Years* (Dallas: Heritage Press, 1993); Steve Coll, *The Deal of the Century: The Breakup of AT&T* (New York: Athenaeum, 1986); Lorraine Spurge, *Failure Is Not an Option: How MCI Invented Competition in Telecommunications* (Encino, CA: Spurge Ink!, 1998).

13. It is unclear if AT&T would even want to lower its prices as a profit-maximizing monopolist. Monopolists seek to maximize profits; firms in perfectly competitive environments seek to maximize revenue.

14. It is very interesting to note that MCI did not intend to initially compete in a low-end disruptive manner. Its intention was to build a service that competed head-to-head with AT&T. The only way it could enter the market in the early days was to take on characteristics that made it look like a low-end disruptive innovation. Once the government lifted constraints, MCI quickly moved to replicate AT&T's business model. Still, in its early days MCI gave an underperforming product to price-sensitive customers.

15. Nonsanctioned equipment (anything not made by Western Electric) was known as a "foreign attachment." Within published tariffs—lists of rates and services with obligations and restrictions on service use—were clauses that subjected customers using foreign attachments to fines.

16. Three years before the Carterfone's introduction, the courts decided to allow customers to use a simple snap-on device introduced by the Hush-a-Phone Corporation. Essentially a rubber cup that shielded background noise, Hush-a-Phone's seemingly innocuous invention caused an incredible stir. AT&T's managers took legal action and approached the FCC. After careful consideration, the FCC decided in 1955 that the device offered by the Hush-a-Phone Corporation would be "deleterious to the telephone system and injure the service rendered by it." However, the courts overturned the FCC decision in 1956 and established the ability of subscribers to use the telephone in "ways which are privately beneficial without being publicly detrimental." See Kevin G. Wilson, *Deregulating Telecom-*

munications: U.S. and Canadian Telecommunications, 1840–1997 (Lanham, MD: Rowan & Littlefield, 2000), 111, 420; quotes from *Hush-a-Phone Corp. v. United States,* 238 F2d 266 (D.C. Cir 1956).

Theory explains why AT&T presciently protested the Hush-a-Phone rubber cup, even if its efforts ultimately were fruitless. AT&T probably realized better than anyone else did that there were other places within its network where competitors could theoretically enter without interfering with the rest of the network. At these modular interfaces, any third-party device that met a set of specifications could plug into the network without any fear of network harm, taking business away from AT&T. AT&T didn't want to let that door open, even a crack.

17. Displacements can throw established firms off balance, because they often require that a firm rethink how it segments its markets. Incumbents that are organized around defined product segments often find it difficult to respond when a new entrant segments the same market differently. For example, the organization of the United States Postal Service and UPS by product categories (first-class mail, etc.) hindered their ability to respond to Federal Express.

18. See, for example, figure 2-4 on page 48 of Christensen and Raynor, *The Innovator's Solution.*

19. This concept was introduced in the appendix to chapter 6 in Christensen and Raynor, *The Innovator's Solution.*

CHAPTER TWO

COMPETITIVE BATTLES

How to Size Up Competitors

When will companies realize the full potential of their innovations and radically change an industry? How can we assess a company's strengths? How can we observe a company's weaknesses? Which of these strengths and weaknesses influence competitive battles? What separates circumstances in which it is prudent to bet on an incumbent versus circumstances in which entrants should have the upper hand? What powers the process of disruption?

The second part of our analytical process shows how to use the theories of innovation to make informed decisions about competitive battles. Taking the tale of the tape and looking for the company with the sword and the shield allow us to answer the competitive battles question: "What is the likely result of head-to-head battles between industry combatants?" Figure 2-1 summarizes the topics covered in this chapter.

Historically, people perceived the process of innovation to be random. *The Innovator's Solution* showed how this seemingly random process actually follows distinct patterns that lead to predictable outcomes. One of the most predictable events is that after establishing an initial foothold, new firms experience a strong incentive to improve, acquire more customers, and migrate into high-profit tiers of their market. This ultimately sets up a battle between the new firms and an entrenched incumbent.

FIGURE 2-1

Competitive Battles

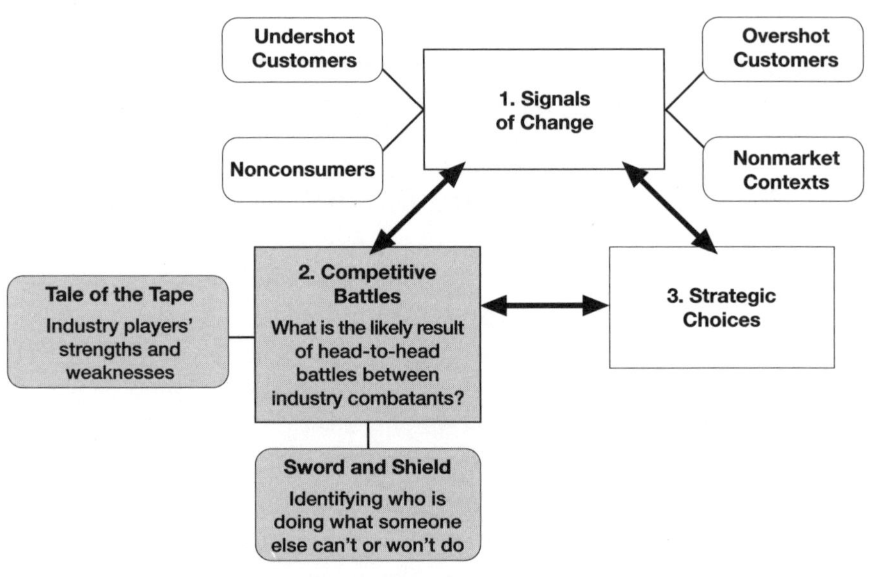

Companies improve their products or services because they want to serve more profitable groups of customers who are willing to pay higher prices for these improved products. Often, it is more than a want. It is a need.

Professor Michael Porter noted that there are two generic strategies for creating a competitive advantage: differentiation and low cost.[1] When a low-cost competitor attacks a given tier of a market, it only enjoys a cost advantage as long as high-cost competitors remain in the market. When a low-cost competitor drives out the high-cost competitors, it must keep reaching further and further up-market to continue to compete against those even higher-cost competitors. Similarly, a company that seeks to create competitive advantage through a differentiation strategy must constantly search for new markets that value the points of differentiation. Only by moving up and into new markets can firms sustain advantages.

People often discount a disruptive company's ability to move up-market. Western Union called Bell's invention a toy. AT&T assumed

MCI was an insignificant blip on its radar screen. Almost all incumbent leaders similarly discounted the ability of their disruptive competitors to move up-market. In fact, one can rarely predict what specific innovations will occur to boost a disruptive firm up-market. What we can always predict, however, is that disruptive firms will be motivated to figure out a way to do it—because it is the path to improved profitability.

Note that this pattern implies a circular process. When companies first start to target nonconsumers, they typically undershoot their needs. So they need to introduce up-market sustaining innovations. Companies trying to meet their customers' needs eventually overshoot their customers, creating opportunities for low-end disruptions and displacements that change the basis of competition. Whereas low-end disruptions and displacements might be good enough for less demanding customers, they will not be good enough for more demanding customers. So they need to introduce up-market sustaining innovations. And so on.

As this cycle unfolds, companies continuously improve their products and services to reach customers willing to pay attractive prices. As they follow their improvement trajectory, they inch closer and closer to the valued customers of established firms. The critical questions are: How will the incumbents react to this attack? Will they flee? Ignore it? Fight to retain their customers? And if they choose to fight, will they win?

Answering these questions involves interpreting the tale of the tape to understand competitors' strengths and weaknesses and looking for the firm with the sword and shield, doing what its competitor cannot or will not do.

Tale of the Tape: Evaluating Strengths and Weaknesses

Firms eventually face competition. Predicting who will win these battles requires evaluating combatants' strengths and weaknesses. Although we have written about this topic numerous times, our goal here is to show how an outside analyst can identify the critical strengths and vulnerabilities that typically aren't visible—even to those who work within a company.

Let's say your friend asks you who you think will win Saturday night's big prizefight. You have never heard of either boxer. What do you do? You quickly flip to a chart in the newspaper that compares the two boxers' heights, weights, historical records, bicep sizes, and so on. This tale-of-the-tape chart gives you a reasonably good sense of each boxer's strengths and weaknesses.

We need a similar way to take a firm's tale of the tape. Most people start and end with what is most visible and obvious: a firm's resources. They see a company's fixed assets, its technology, its cash reserves, its product lines, and the résumés of its management team and think that possession of superior resources determines success.

The RPV theory discussed in the introduction provides a more comprehensive view of what comprises a firm's strengths and weaknesses. Remember, the theory holds that beyond a company's resources, its processes and values define its strengths. Firms succeed at opportunities that fit its resources, processes, and values and struggle with opportunities that run counter to those strengths.[2]

Assessing a firm's RPV requires answering three questions:

1. Does a firm have or can it marshal the *resources* required to attack an opportunity?
2. Do the firm's *processes* effectively and efficiently facilitate it doing what needs to be done?
3. Do the firm's *values* allow it to prioritize one opportunity over other options on its plate?

How can you identify a company's RPV? Resources tend to be visible. Imagining tough problems companies must repeatedly solve can highlight core processes. A firm's income statement and its past investment history provide important clues to its values. Table 2-1 summarizes how to take the tale of the tape and evaluate a firm's RPV.

Identifying Visible and Invisible Resources

Identifying resources tends to be easy. Many resources are visible, such as a firm's technology, products, or cash coffers. It is also easy to surmise a firm's less visible resources: its human capital, accumulated knowledge, or established brand. Consider this assessment of Western Union's strengths. Western Union's resources included "well over $40 million in capitalization . . . [and] $3 million a year in net profits from its busy nationwide network of wires and offices in nearly every city, town, and established settlement in the United States."[3] BellSouth, one of the local telephony companies created in the breakup of AT&T in the early 1980s, had close to one hundred thousand employees, almost 14 million

TABLE 2-1

How to Take the Tale of the Tape

Term	Definition	What to Look For
Resources	Things company has or has access to	• Tangible assets: Technology, products, balance sheet, equipment, distribution network • Intangible assets: Human capital (employees' backgrounds, accumulated skills), brands, accumulated knowledge
Processes	Ways of doing business (skills)	• Difficult problems we know the company has repeatedly solved over time • Typical processes: Hiring and training, product development, manufacturing, planning and budgeting, market research, resource allocation
Values	Prioritization determinant (motivation)	• Business model: – Way a company makes money (e.g., mix of sales revenue to postsales service revenue) – Cost structure/income statement – Size and growth expectations • History of investment decisions—what has been prioritized in the past

access lines, more than $10 billion in revenues, more than $20 billion in assets, millions of miles of copper wire to consumers' homes, and licenses to develop wireless technologies in Atlanta, Miami, and New Orleans. All of these were valuable resources.

Resources are quite flexible. They are *things* that firms can acquire and build or sell and destroy. The same resource can be used productively in a number of different markets or organizations. One final note about resources: It is not so much that a firm *has* a resource. It is that a firm *has access* to a resource. Successful entrepreneurs have the ability to tap into resources beyond their immediate control.[4]

Identifying Processes by Looking for the Tough Problems

In contrast to assessing a firm's resources, it is a much more difficult task for an outside analyst to understand the nature of a firm's processes. Processes are the patterns of interaction, coordination, communication, and decision making employees use to transform inputs of resources into products, services, and other resources of greater worth.[5] When companies must solve the same problem repeatedly, they develop formal and informal processes so they can successfully address the task every time it arises, minimizing the risk of failure. If they didn't develop and use processes to address recurrent tasks, they would constantly waste time reinventing solutions. Indeed, the term *organizational learning* in many ways consists of fostering and following processes.

Unlike resources, processes are *meant* not to change or, if they must change, to change through tightly controlled procedures.[6] So processes define skills and strengths, and skills and strengths define inabilities and weaknesses. When people use a process to do the task it was designed for, it is likely to do the task effectively. When people try to use that same process for a very different task, it often seems highly bureaucratic and inefficient.

For example, in the introduction we discussed Microsoft's software development process. Microsoft must have a robust process to manage the development of its next-generation operating system, or else absolute chaos would ensue. Microsoft breaks each software program into modules and then repeatedly compiles an overall program throughout its development. These processes allow Microsoft to manage a very complicated task over and over again. If Microsoft tried to use the same process to manage the development of a small application that required quick customization and turnaround time, the process would be a liability.

The best way to determine a company's processes from the outside is to imagine the kinds of problems a company must have repeatedly solved that have led to its success. Phone companies have to build and maintain large, complicated telecommunications networks that must work just about all the time. Airplane manufacturers Boeing and Airbus have to coordinate a complicated network of suppliers. Johnson & Johnson has to gain approval for new medical devices. Procter & Gamble has to develop product marketing plans. For these companies to be successful, they must have developed ways to solve these problems again and again. They need processes that facilitate their ability to get done what has to get done.

There also are less visible background processes that support critical decisions such as where to invest resources, how market research is habitually done, how financial projections are created, how plans and budgets are negotiated internally, and so on. Many of these important processes that define a firm's strengths are not readily observable to outsiders (or, as a matter of fact, to insiders). What we are saying, however, is that a listing of the recurrent problems or tasks that a company has successfully, repeatedly addressed is a visible and reasonably accurate proxy for a listing of its processes. This means that if a company has never confronted a particular problem or task before, an optimized process to complete that task would not exist.

Identifying Values by Looking at Income Statements and Past Investment Decisions

We attach a particular meaning to the word *values*. Values do not refer solely to what a company deems to be ethically acceptable. An organization's values are the criteria employees use when making prioritization decisions. Every employee at every level of the corporation makes these sorts of decisions every day. For example, salespeople have to decide whether they should call customer *A* or customer *B*. Once they decide whom to call, they must make an on-the-spot decision about which products to push versus which products to de-emphasize as they interact with customers and discuss their priorities. Engineers decide which project they will work on today and which project they will tackle tomorrow. They make *this* design choice and not *that* one.

Values also determine the larger strategic decisions that senior management makes. Do we acquire this company or that company? Do we grant the budget request of this business unit but cut the budget of that one? Values are the criteria that drive a company's resource-allocation process—the mechanism that defines which threats and opportunities the company will pursue, and which it will not.

How can you identify these values? By looking for proxies. A firm's revenue mix, cost structure, absolute size, most important customers, and history of past investment decisions can help you understand the sorts of strategies and investments that will appear profitable to its managers, and which will appear unattractive. It is a pretty safe bet that managers will accord highest priority to the opportunities that are financially most attractive.

The first place to look to identify a company's values is its income statement. What is its revenue mix? Does it earn a significant proportion of its revenue from many products or a specific subset of products? From postsales service? A company is always unlikely to prioritize opportunities that destroy significant revenue streams. What are the gross profit margins it needs to earn to support its cost structure? A firm that has a cost structure that requires a 60 percent gross margin is typically not interested in opportunities with a 20 percent gross margin. How big does a new opportunity have to be to matter to the firm? An opportunity that is attractive to a firm that has $50 million in sales and seeks 10 percent top-line growth would be unattractive to a firm that has $5 billion in sales and seeks 10 percent top-line growth. A $2.5 million opportunity meets 50 percent of the first firm's growth needs but only 0.5 percent of the second firm's growth needs. Which company do you think will place a higher priority on going after a $2.5 million market?

Western Union's income statement and annual report pointed to the importance it placed on long-distance data communications, which made up almost all of its revenue. Its 1869 annual report detailed how all managers had to report the "number of messages sent and received, the gross receipts, the amounts received on messages for each office with which business was done; the amounts received at all other offices with which messages were exchanged; the amounts received for or paid to other lines, and all expenditures in detail."[7] Managers were measured on these variables and cared a great deal about them.

The next clue to a company's values comes from its customer roster. A company needs to prioritize innovations that improve its ability to serve its most important customers. Companies that derive a high proportion of their income from a certain class of customers are likely to focus on innovations that target those customers. Who were Western Union's leading customers? Companies such as railroads, newspapers, and financial brokerage houses. These companies all greatly valued long-distance, real-time communications, especially Western Union's continuous data services, where they could receive uninterrupted reporting from commodity and stock exchanges. In particular, the telegraph played a critical role in the management and coordination of the railroads.

The final way to see a firm's values is to look at its history of past investment decisions. Which opportunities did it decide to target and which opportunities did it decide to forgo? The stream of incremental investment decisions Western Union made in the 1860s and 1870s indicated that its overwhelming focus was on its long-distance data offering. Its history of investment showed that its values prioritized investing in innovations that enhanced its core offering.

Taking the tale of the tape and identifying a company's resources, processes, and values provides much deeper insight into what a firm can and cannot do. The next section shows how comparing competitors' RPVs and looking for important differences—what we call *asymmetries*—provide a way to evaluate head-to-head battles.

Looking for the Sword and the Shield

When companies have the same capabilities and motivation, they care about the battle and have the necessary skills to fight it. Skills in execution make the difference here—and because other scholars have addressed these challenges quite capably, we do not focus on them in this book.[8]

The more interesting scenarios occur when there are asymmetries—important differences of motivation or skills. Asymmetries of motivation occur when one firm wants to do something that another firm specifically does *not* want to do. Asymmetries of skills occur when one firm's strength is another firm's weakness.

The section discusses three topics:

1. How asymmetries power the process of disruption
2. How to identify the company with the shield of asymmetric motivation and the sword of asymmetric skills on its side
3. How to identify circumstances in which a high-potential disruptive development will prove disappointing, ending in either a brutal fight or incumbent co-option

How Asymmetries Propel Disruptive Entrants

Asymmetries allow disruptive attackers to enter a market, grow without incumbent interference, and mitigate the incumbent's response when it is finally motivated to counterattack. The result of asymmetric battles often is the seemingly sudden end of a great firm. From the incumbent's perspective, every action it takes is rational. But the outcome is devastating.

Disruption is the strategy that creates and capitalizes on asymmetries of motivation and skills. It naturally follows a three-step process:

Step 1: Entrants enter behind a shield of asymmetric motivation; early incumbent response leads to cramming. Incumbents pass over what in retrospect turn out to be multibillion-dollar opportunities because attackers take advantage of asymmetries of motivation. When people say "flying beneath the radar," they really mean "taking advantage of asymmetries of motivation." Disruptive markets start among customers that appear to the incumbent to be either undesirable or nonexistent. The initial absolute size of a disruptive opportunity is generally too small to justify any substantial amount of investment or even management attention.

Asymmetric motivation shields companies from competitive response, because their potential challengers are just not interested in fighting. Even if they fight, their hearts aren't in it. The battle matters more to their opponents. Remember the old saying: "It is not the size of the dog in the fight. It is the size of the fight in the dog."

Gauging asymmetries is so essential to predicting outcomes because the availability of information about a threat or opportunity has little influence on who wins and who loses. What makes the difference is what a company *does* with that information. Incumbents usually see the same technologies that entrants do. Because of their processes and values, however, incumbents predictably "cram" the technology into the largest and most obvious market applications.

What causes cramming? Managers tend to adopt innovations in ways that make sense to their businesses. They know the markets their companies serve. Those markets appear large and measurable. The companies have established processes to help them penetrate that market. So the established competitors in a market naturally choose to bring all innovations—sustaining and disruptive—to their core markets where their best customers reside.[9] For example, had Western Union purchased Bell's patents, we would predict that it would not have commercialized the technology until it had toiled away in its labs modifying the technology so it could be useful to Western Union's most profitable real-time long-distance data customers.

When an existing firm tries to insert a product or service with disruptive potential into its processes, what comes out the other end tends not to be disruptive. Instead of embracing the innovative product's inherently disruptive nature, the incumbent inevitably tries to morph the product to fit into its existing processes and values. It alters the innovation to enhance its appeal to core customers and fit within its operating model. The problem with cramming is that it changes the innovation in ways that obviate its inherent disruptive energy. It takes an innovation from a circumstance in which its unique features are valuable to a circumstance in which its unique features are a liability.

Cramming (displayed in figure 2-2) is like trying to stuff a square peg into a round hole. What signs indicate that cramming is occurring? When companies spend a lot of money fixing product deficiencies, they may be cramming. Large charges or expenses to integrate an acquisition are a good tip-off. Another sign is when companies must convince customers to change their behavior or put up with something they don't seem to want.

For example, Kodak first began to sense that digital imaging might pose a threat to its core business in the mid-1990s. It invested more than $2 billion in research and development. However, it

FIGURE 2-2

Cramming

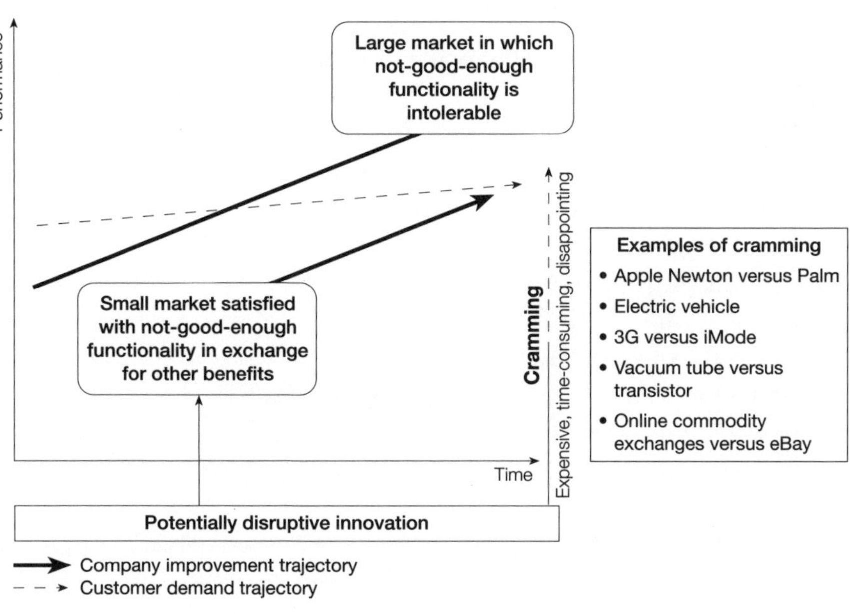

framed the challenge as, "How do we make digital imaging good enough to serve as a viable replacement to silver halide film in our core market?" By seeking to create high-priced, performance-competitive digital products, Kodak missed much of the disruptive growth driven by inexpensive digital imaging. Kodak eventually established a strong market share after introducing a very low-cost camera, but only after spending $2 billion trying to maximize its cameras' performance.[10] Subsequent chapters present other classic cramming case studies, describing how vacuum tube manufacturers crammed the transistor (chapter 7) and how incumbent wireless providers crammed next-generation wireless data services (chapter 10).

Cramming happens all the time. It is almost *never* successful. It costs a lot of money and usually ends with disappointing results. Cramming explains why so many disruptive innovations originate from within incumbents but are ultimately commercialized by separate organizations. When a firm develops the germ of a disruptive idea, it often falls prey to the temptation to cram that opportunity to its mainstream market. The momentum behind the widely practiced mantra "serve our customers" is often nearly impossible to overcome. The mainstream market tends to reject the innovation. Frustrated managers and engineers leave, form a new company, and discover a new market where the innovation has value.

Step 2: Entrants grow and improve; incumbents choose flight. As disruptive attackers follow their own sustaining trajectories, they make inroads into the low end of the market or begin pulling less demanding customers into a new context of use. What happens when the disruptive entrant begins to make inroads? A good way to visualize what incumbents can do when faced with a disruptive attack is to consider how humans respond to a perceived threat. Our body immediately reacts. We produce adrenaline. Our heart rate goes up. Our respiration rate goes up. Blood flow redirects from nonessential areas to critical areas. Our body is prepared for one of two actions: fight or flight.

Incumbents naturally choose flight. What looks highly attractive to the entrant continues to look relatively unattractive to the incumbent. The asymmetric motivation leads to incumbents naturally fleeing the low end. They cede that market to the entrant. AT&T, for example, initially ceded the lowest end of the existing long-distance market to MCI. Western Union clearly ceded the new local communications market to the Bell companies.

Remember, incumbents focus on delivering up-market sustaining innovations that allow them to earn premium prices by reaching undershot customers. They view flight as a *positive* development. When there are large groups of undershot customers in the higher tiers of a market, incumbents can beat a long and profitable up-market retreat. The customers incumbents leave behind tend to be the least attractive. Sales can go up (high price points replace low price points). Margins typically go way up. The incumbent stops worrying about disloyal, dissatisfied, low-paying (overshot) customers whom outsiders may term "distractions" and instead focuses on very loyal, highly satisfied, high-paying customers. In essence, incumbents can disassociate themselves from a business in exchange for a better one.

How can we identify when flight is in progress? Changes in customer or product mix can be a clear sign of flight, as are plans to discontinue low-end product lines or to stop servicing old versions of products. Companies tend to announce that they are "focusing on the core" or "seeking higher margin opportunities" when flight is in progress. Sometimes companies will launch diversification efforts to flee from an attacker.

Step 3: Entrants utilize the sword of asymmetric skills. What starts small gets big—too big for an incumbent to ignore. The number of undershot customers dwindles. But when incumbents become cornered, they face two problems. First, asymmetric motivation still stymies effective response. Even though the new opportunity may appear big, it typically requires a different business model. Even worse, incumbents are now at the mercy of asymmetric skills. Remember, disruptive innovations typically introduce new benefits to a market, usually centered on convenience, simplicity, customization, or affordability. As the entrant steadily solves unique problems, it builds the ability to do whatever is required to succeed in its context. When the incumbent has retreated into the highest tiers of its market and *has* to fight because there is no room for further retreat, it is at a competitive disadvantage. As the game changes to the one the disruptor plays best, it is *very* hard for the incumbents to develop new skills quickly.

By the time the Bell companies firmly established themselves, they developed unique competencies related to transmitting the human voice over relatively short distances. They established skills in acoustics, network management, customer service, and so on. Western Union had none of these skills. Its business did not need to solve these problems. It was on the wrong side of asymmetries. Western Union couldn't suddenly become a viable competitor after the telephone had been improving for twenty-five years. Similarly, Digital Equipment Corporation couldn't match the flexibility of the personal computer assemblers' processes, Sears couldn't match the inventory turns and low prices of the discount retailers, and so on.

It is important to note that almost thirty years elapsed from the introduction of the telephone before telephony operators began making serious inroads against Western Union. Similarly, wireless telephony existed for twenty-five years before it seriously began to erode the wireline business. Both innovations grew for a very long time in markets that were different from the incumbents' cores. Incumbent firms that take action when data shows a downturn in their core businesses take action too late. The only signal to take timely action is sound theory.

In disruptive circumstances, entrants win because they are—often unknowingly—taking advantage of the asymmetries on their side. An incumbent's strengths are its weaknesses. Its values, which make sure it delivers excellent products to demanding customers, stop it from going after markets where ultimately its strongest competitors will forge their processes and values. The incumbent's processes, those that allow it to serve its customers well, are weaknesses when the game changes and new capabilities are necessary. Fleeing from the disruptive attacker feels good in the short term but further deprives the incumbent of the necessary skills to compete. The end can come swiftly and can appear stunning to the untrained eye. Typically, the best an incumbent can do is to belatedly acquire the winning firm and stave off ultimate destruction.

In sustaining circumstances, incumbents win because *they* are the ones attacking with asymmetries on *their* side. Consider an entrant that tries to bring a radical sustaining innovation to an incumbent's best customers. There are no asymmetric motivations here. Incumbents are *very* motivated to go after this market. The incumbent's processes are relative strengths, not relative weaknesses. A new firm that attempts to introduce a radical sustaining innovation almost guarantees itself a long, bruising fight. Unless the new firms sell their advanced product quickly to the established leaders or have extremely patient, deep-pocketed investors, incumbents will spend more, produce better products and services, and ultimately force the entrants out of the market.[11]

Identifying the Firm with the Sword and the Shield of Asymmetries

The power of asymmetries is why the signals of change question is so important. Companies that develop different ways to target nonconsumers and overshot customers can create a new market or attack the lower tiers of a market almost free from interference from an incumbent that views the opportunities as unattractive. They have the potential to develop legitimately different skills and business models.

Table 2-2 shows how to identify the two types of asymmetries that help to determine winners and losers.

There are three factors that contribute to asymmetric motivation, all of which relate to firm values. The first factor relates to an opportunity's absolute size. An opportunity that looks interesting and large to a small firm might look uninteresting and small to a big firm. To the emerging

TABLE 2-2

Identifying Asymmetries of Motivation and Skills

What to Look For	Definition	Signals
Asymmetric motivation	Firm does something that another firm does not want to do (provides shield protecting from response)	• Size of market relative to firm size • Target customers • Business model in market relative to existing business models
Asymmetric skills	Firm does something another firm is incapable of doing (provides sword to use during attack)	• Mismatch between processes required for success and established processes

telephony companies, the emerging local communications market looked boundless. Through the lenses of Western Union's values, the market looked inconsequential. Similarly, a meaningful growth opportunity for a *Fortune* 50 company will be very different than that of a start-up. The second factor relates to an opportunity's initial customer. Disruptive markets start among customers that appear to the incumbent to be either undesirable or nonexistent. Entrants are motivated to serve the very customers incumbents are motivated not to serve. The final factor relates to an opportunity's business model. Disruptive entrants use business models that do not fit the ways established firms make money. Gross margin per unit sold tends to be lower but turnover or asset utilization tends to be higher. Disruptive innovations tend to be off-the-shelf products, in which the customer turns either to a group of specialist firms or to themselves to provide postsales service. A company that has a business model based on long-term relationships and multiyear support agreements will have little interest in selling a product that obliterates those revenue streams.

How can you observe asymmetric motivation in action? When companies take completely different actions that make sense to both of them, it is a sign that there are asymmetries. When one firm calls an industry "unprofitable" while another firm calls that market "important," asymmetries are at work. For example, in the 1990s, Cisco raced to capture more of the networking equipment market while IBM sought to leave networking equipment to focus on higher-margin services. On the other hand, when an incumbent announces an entrant's emergent growth market is a strategic priority, it could indicate a lack of asymmetries.

Asymmetric skills act as a weapon a company can brandish to attack its opponents. Remember, a company's skills come largely from its processes. A process comes from repeatedly solving a particular class of problem. Processes are designed to get the same thing done, over and over—and as such they tend to be inflexible. Asymmetric skills arise when one firm, through repeatedly completing the same task, has developed a unique ability to do something that its competitor is uniquely unable to do.

How can you tell if combatants have asymmetric skills? Make a list of the tasks the company has repeatedly addressed, for which formal and informal processes have likely coalesced. Compare the list to the nature of tasks required to succeed in the disruptive market. If a company's processes facilitate its doing what it needs to get done in a market context, it has the requisite skills to take that market. Companies all have strengths and weaknesses. When one firm has strengths in markets in which another firm's capabilities are weaknesses, the firms have asymmetric skills.

Circumstances in Which High-Potential Disruptive Development Will Prove Disappointing

There are certain situations in which—even though disruptive entrants seem to be on an unencumbered trajectory toward success—incumbents still emerge triumphant.

Understanding asymmetries allows us to identify two specific circumstances in which the disruptive process is likely to come off the rails. The first circumstance occurs when the industry context makes incumbent flight unpalatable. The second occurs when an entrant fails to develop a distinctive business model or to create unique skills in its early stages, making *co-option* the natural choice for the incumbent. In these circumstances, expect to see tough competitive battles after the aggressive growth phase.

Industry Contexts in Which Flight Is Unpalatable

Fleeing up-market is only a natural choice for incumbents when there is an adequately sized, attractive market at the high end. This isn't always the case. Firms might not have the ability to reach the next tier of undershot customers. Or undershot customers might not exist. Or an incumbent's cost structure and business model sometimes prevent it from leaving low-end markets.

Think about the battle between AT&T and MCI. MCI grew without real interference from AT&T for a few years. But AT&T eventually *had* to respond to MCI. AT&T's business model depended on amortizing large fixed costs over a wide number of users. Flight to the high end—which typically entails focusing on smaller numbers of customers who are willing to pay high prices for the most advanced products and services—really wasn't a competitive option. Because AT&T cared about the customers MCI took from it, it ultimately decided to fight back. The resulting "long-distance wars" of the 1980s and 1990s cost both companies millions in advertising, with the "winner" capturing a bigger slice of an increasingly less lucrative market.

When firms *can't* choose to flee, we can expect them to fight. Entrants can no longer hide behind asymmetric motivation. And if this happens early enough, entrants don't have the time to refine their asymmetric skills. The result is often a bruising battle for market share. This concept helps explain the no-win situation facing many players in the aviation industry (discussed in greater detail in chapter 6).

Lack of Fully Developed Asymmetries Make Co-option a Natural Choice

Incumbents can't flee when they have no up-market to penetrate. When else are incumbents strongly motivated to respond to disruptive attackers? When an entrant makes money in similar ways or uses similar processes to incumbents. An entrant runs into trouble if it does not develop a business model that is unattractive to the incumbent, or if it does not hone unique skills that are matched to the disruptive business. It can create *initial growth* by taking advantage of the incumbent's disinterest in small, immeasurable markets. But once the entrant has proved to the world that the market exists, the incumbent can muster its internal resources to co-opt the innovation, unless the entrant has made that path unattractive. Natural incumbent motivation shifts from flight to fight. And in direct fights with comparable motivation to win, the incumbents have some real advantages to bring to the battle.

What indicates an attempt at co-option? One sign is an effort by incumbents to develop products or services that mimic the entrant's offering. Acquiring and attempting to integrate one of the class of disruptive entrants is a clear signal that the incumbent is attempting co-option. When business model or process asymmetries don't exist, co-option becomes a viable incumbent response strategy.

Incumbents might still miss much of the growth created by the disruptive innovation, of course, depending on *when* they initiate following a co-option strategy and *where* they implement it. *Growth-driven* co-option needs to occur early and entails going after an entrant's core customers. In essence, an incumbent co-opts the disruption by augmenting its existing product and service offerings to appeal to new customers.[12] When confronted with wireless technologies, for example, the incumbent phone companies chose growth-driven co-option. They developed wireless technology using internal resources and tried to market it to new customers.

Defensive co-option usually occurs later in a technology's development. Incumbents recognize that they have lost the game in the volume end of the market and do what they can to block incursions from below. For example, Oracle introduced its disruptive relational database in the minicomputer market in the 1980s.[13] When IBM realized that Oracle had decisively won the game in that market, it introduced a relational database at the low end of its mainframe market, attempting to block Oracle's advance.

Look to target customers and company announcements to see if companies are following growth-driven or defensive strategies. For instance, a company announcing that the disruptor's home market is a strategic priority is a clear signal of a growth-driven response.

Our theories suggest co-option efforts will be fruitless when success requires different skills or motivation. But when co-option is viable, incumbents can still ultimately master the technology and detain or swallow potential attackers. Existing telephone companies knew how to build and maintain networks. They had existing customers. They had a ton of cash. These were tough capabilities for wireless entrants to overcome. Selecting a business model that looked attractive to incumbents and providing a means for incumbents to learn put wireless entrants at a long-term disadvantage.

To sum up, incumbents tend to respond to potential disruptive incursions in one of three ways: they either cede a market, attempt growth-driven co-option, or attempt defensive co-option. Table 2-3 summarizes how to identify these response strategies.

The Principle of Relativity

The concepts discussed in this chapter raise one other important point: that disruptiveness is a relative concept. A firm's resources, processes, and values filter the way a firm looks at an innovation. Firms with different resources, processes, and values can look at the same innovation and see very different things. For instance, some firms looked at the Internet and saw a nifty way to lower internal costs. Other firms looked at the Internet and saw a vehicle capable of driving new growth.

This principle of relativity explains a seemingly paradoxical statement: A firm cannot be disruptive without being sustaining. Once a firm

TABLE 2-3

Identifying Incumbent Response Strategies

Strategy	Definition	Signals
Ceding	An incumbent leaving a market to an entrant	• Company announcing a refocusing on core customers • Abandoning lower-tier markets • Plans to discontinue low-end products
Co-opting	An incumbent attempting to fight an attack using internal resources	• Company building or acquiring disruptive innovations
Growth-driven	An incumbent targeting the entrant's customers	• Incumbent targeting entrant's market with modified version of core product • Announcements that entrant's market is a strategic priority
Defensive	An incumbent building a wall around its existing customers to prevent entrant entry	• Incumbent bringing new product to low end of existing customer base • Announcements that entrant's market is not a priority

establishes its disruptive foothold, subsequent innovations move it up along its own improvement trajectory. These additional innovations are sustaining relative to the prior market position of the disruptor. But they are highly disruptive relative to the established firm because they allow the attacker to get good enough to meet the needs of ever-larger groups of customers, setting up the eventual competitive battles discussed in this chapter.

The second part of our approach involves assessing competitive battles. Sizing up a head-to-head competition first involves assessing the combatants' strengths and weaknesses. Just as boxers can be evaluated using measures such as their reaches and records, firms can be evaluated by looking at their resources (what they have), processes (how they do their work), and values (what they want to do). Taking the combatants' tale of the tape identifies their strengths and weaknesses, their capabilities and motivations.

The winners of the battles we care about are doing something another firm either does not want to do or cannot do. These asymmetries power the natural course of disruption. New companies develop new ways to target nonconsumers or overshot customers. They build new growth hiding behind a shield of asymmetric motivation. Incumbents just don't want to respond. As entrants grow, they forge their sword of asymmetric skills by developing unique skills to provide their unique offerings. When incumbents want to respond, they are no longer able.

Key questions to ask when analyzing competitive battles include:

- *What are industry players' business models? What are their motivations? What are their skills?*
- *How do industry players compare to one another? How do they compare to the needs of the market? Where are there symmetries? Where are there asymmetries?*
- *Do the asymmetries tilt in favor of the attacker or the incumbent?*
- *Does the innovation naturally fit its target market? Is there evidence of cramming?*
- *Are there signs that a company is ceding a low-end market and trying to move up? Is there an "up" to move to? For how long?*

Evaluating these two topics raises important questions. Where do asymmetries come from? What choices can firms make to enhance their chances of success? What choices can lead firms to fail? The next chapter shows the strategic choices to analyze to help answer these sorts of questions. It shows the decisions firms can make to harness the forces that act upon them to change the course of events.

Notes

1. See Michael E. Porter, *Competitive Strategy: Techniques for Analyzing Industries and Competitors* (New York: Free Press, 1980), and *Competitive Advantage: Creating and Sustaining Superior Performance* (New York: Free Press, 1985).

2. An excellent study of how resources, processes, and values drive outcomes is Toma Noda and Joseph L. Bower, "Strategy Making as Iterated Processes of Resource Allocation," *Strategic Management Journal* 17 (1996):

159–192. This study showed how different processes and values led local telephone companies with similar resources to commercialize wireless technologies in very different ways.

3. George David Smith, *The Anatomy of a Business Strategy: Bell, Western Electric, and the Origins of the American Telephone Industry* (Baltimore: Johns Hopkins University Press, 1985), 36.

4. For an excellent handbook that addresses how entrepreneurs can find success without directly controlling assets, see William A. Sahlman, Howard H. Stevenson, Michael J. Roberts, and Amar V. Bhide, *The Entrepreneurial Venture* (Boston: Harvard Business School Press, 1999). Of course, the VCE theory suggests there are some assets a firm *must* have direct control over if it is to capture value.

5. For more on processes see David Garvin, "The Processes of Organization and Management," *MIT Sloan Management Review* 39, no. 4 (summer 1998): 33–50.

6. Some commentators believe that processes can be designed to be flexible. Our opinion is that a good process is by definition inflexible—it is designed to do the same task well, over and over again. Getting that thing done well means that there are other things it cannot do well. Often when people think about flexible processes, they really mean having multiple sets of processes with some kind of mechanism to say that task *A* goes to process 1 and task *B* goes to process 2. We know of no processes that work well at everything. Every process works poorly at some tasks.

7. Alfred D. Chandler Jr., *The Visible Hand: The Managerial Revolution in American Business* (Cambridge, MA: Belknap Press of Harvard University Press, 1977), 198.

8. See, for example: Larry Bossidy and Ram Charan, *Execution: The Discipline of Getting Things Done* (New York: Crown Business, 2002); James C. Collins and Jerry I. Porras, *Built to Last: Successful Habits of Visionary Companies* (New York: HarperBusiness, 1994); James C. Collins, *Good to Great: Why Some Companies Make the Leap . . . and Others Don't* (New York: HarperBusiness, 2001); William Joyce, Nitin Nohria, and Bruce Roberson, *What (Really) Works* (New York: HarperBusiness, 2003); Steven C. Wheelwright and Kim B. Clark, *Revolutionizing Product Development: Quantum Leaps in Speed, Efficiency and Quality* (New York: Free Press, 1992).

9. Of course, it is worse than this. Not only does an incumbent try to bring the innovation to its existing customers, it typically tries to bring it to its *best* existing customers. Ironically, these customers value the new attributes of the disruptive innovation the least.

10. For more on this, see Clark Gilbert and Joseph L. Bower, "Disruptive Change: When Trying Harder Is Part of the Problem," *Harvard Business Review*, May 2002, 94–101; and Clark Gilbert, "Can Competing Frames Co-exist? The Paradox of Threatened Response," working paper 02-056, Harvard Business School, Boston, 2002.

11. The best strategy for entrants in this situation is to try to briefly leap in front of incumbent companies and then sell themselves to incumbents. Many venture capital firms, whose partners have less and less tolerance for taking market risk, are adopting this strategy. They fund companies that are developing a sustaining innovation and seek to sell the start-up to an incumbent for a quick profit. Airbus is a notable exception that proves the rule here (discussed in more detail in chapter 6). Airbus has pursued Boeing with a sustaining innovation strategy—but survival and success have required European governments to cover tens of billions in losses.

Entrants can undoubtedly move more quickly than established firms can. But once established firms marshal the resources to respond, entrants are in trouble. Other researchers have noted that some industries teeter on a razor's edge. These industries are so precarious that companies have no margin for error. In these industries, entrants can triumph by introducing better products, rapidly locking in customers, and quickly moving up the experience curve. Rebecca Henderson noted this phenomenon in her doctoral thesis, in which she studied the photolithographic alignment equipment industry. In four successive generations, new firms ended up toppling established market leaders with what we would classify as sustaining innovations. Our belief is that these industry circumstances do exist but are quite rare. See Rebecca M. Henderson and Kim B. Clark, "Architectural Innovation: The Reconfiguration of Existing Systems and the Failure of Established Firms," *Administrative Science Quarterly* 35 (1990): 9–30.

12. Again, remember that we mean incumbents to be firms already entrenched in a market context. We would classify an existing firm launching a disruptive attack on a new market as an entrant.

13. *Hierarchical* database software was designed for use on mainframe computers in the late 1960s and 1970s. Data in hierarchical databases is organized in a "tree" format, like a classic organization chart. These databases required expert programmers but could perform specific searches very quickly and used the mainframe's resources efficiently. The databases were ideal for demanding customers who needed heavy-duty transaction processing, such as financial institutions. The leaders in the industry were companies such as IBM and Cullinet, which in 1978 became the first U.S. software company to go public. In the early 1970s, a researcher at IBM came up with the theory behind a new form of database, called a *relational* database. All pieces of data in a relational database are interrelated. The initial relational databases were slow and used a lot of computer resources, but allowed users to relatively quickly and easily run customized queries without having expert knowledge about the underlying database structure. This technology initially had little value to mainstream customers because its functionality was not good enough. With their relative simplicity and flexibility, relational databases were clearly more convenient than hierarchical databases.

CHAPTER THREE

STRATEGIC CHOICES

Identifying Which Choices Matter

Is disruption inevitable? What are the important decisions facing entrants? What initial market should they target? How should they design their organizations? What factors influence these sorts of decisions? How can an outsider observe them? What response strategies allow incumbents to fend off legitimate disruptive attacks? How can we evaluate the likely success of these response strategies?

The third part of our analytical process shows how to identify the important choices a firm makes and understand their implications. Assessing an entrant's preparation regimen and its value network selection and evaluating whether an incumbent has earned its disruptive black belt allow us to answer the strategic choices question: "Are firms making decisions that increase or decrease their ultimate chance of success?" Figure 3-1 summarizes the topics covered in this chapter.

The theories of innovation explain the natural forces that affect every firm. The forces typically manifest themselves in subtle ways, in the day-to-day decisions managers make. Existing firms typically succumb to those forces without even realizing it, because they make what appear to them to be correct, rational decisions.

FIGURE 3-1

Strategic Choices

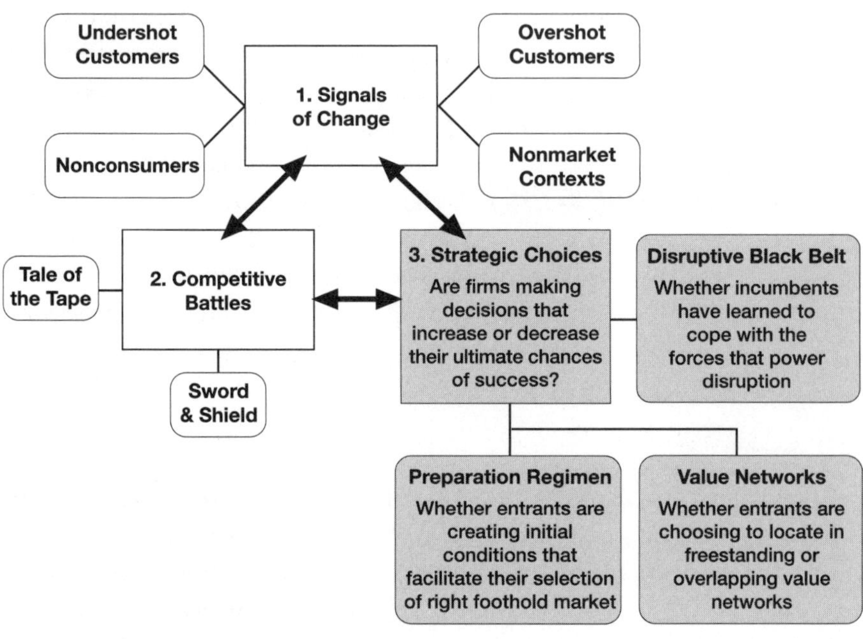

But the game does not have to end in despair. Firms are complicated entities. The path of disruption has many twists and turns. The winner of a competitive battle is rarely preordained. Whether an innovation is deployed in a disruptive or a sustaining way is typically a management decision. Companies can create advantages, and companies can counter advantages. Put simply, *choices matter*.

This chapter discusses three critical choices that can alter the natural course of industry evolution:

1. Entrants following the wrong preparation regimen. A preparation regimen consists of a firm's hiring decisions, its strategy-making processes, and its funding sources. The wrong decisions can create initial conditions that lead entrants to the wrong foothold market.
2. Entrants creating overlapping value networks that provide easy avenues for incumbent co-option. A value network involves up-

stream suppliers, downstream channels to market, and ancillary providers. If an entrant chooses to compete in an incumbent's value network, it leads to pressures to conform to the incumbent's values, thereby erasing the asymmetries that the entrant needs.

3. Incumbents earning their disruptive black belts and developing the ability to master the forces that act upon them. Just as airplane pilots with the right equipment and skills can fly despite the pull of gravity, firms can master disruption.

Any of these three choices can flip the advantage back to incumbents who can leverage their often superior resources and established processes.

The Wrong Preparation Regimens Often Lead to the Wrong Footholds

Finding the wrong entry point or foothold can quickly put entrants on the wrong side of asymmetries. From day one, every entrant faces a tempting choice. Developing completely new products is hard. Finding new markets is harder. Identifying large, existing markets is easy. The most demanding customers in those markets tend to pay the highest prices. Trying to reach these customers involves following sustaining, not disruptive strategies. The previous chapter showed why incumbent firms have a leg up in battles based on sustaining innovations. It showed what happens when companies try to cram potentially disruptive innovations into existing markets.

Strategic Choice: Wrong Preparation Regimen Leads to Selecting Wrong Foothold

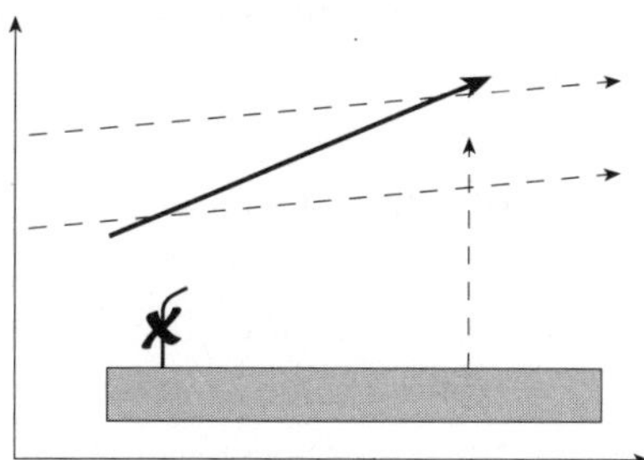

Of course, you can observe these sorts of developments. But is there any way to tell beforehand whether entrants are likely to set off down these treacherous paths? Think back to our boxing analogy from the last chapter. If you want a realistic read on a fighter's strengths and weaknesses, you would not trust yourself just to look at this fighter and conclude, "He's very strong," or "She's fast." You'd really need to understand the preparation regimen that the fighter followed to get ready for the fight. That would give you a much better sense of the fighter's likely endurance, reaction times, and so on. The same is true in evaluating a company's strengths and weaknesses. Some of these might be obvious just from looking at the company. But speed, endurance, and patterns of response can only be understood by examining whether and how the company has prepared for the present challenge.

The Innovator's Solution discussed theories to help innovators grapple with important decisions such as setting strategy, hiring, and sourcing investment capital.[1] Here, we show how outsiders can use these theories to *evaluate* whether a firm is using a strategy-making process, has hired the right people, and has sought funding from the right sources in ways that will increase its chances of creating the conditions that facilitate finding the right foothold.

Addressing the preparation regimen issues grows easier the closer one is to the core of a firm. But even those far from the core can gain insight into these topics, if they know what questions to ask.

Table 3-1 summarizes the important components of a firm's preparation regimen, the tool or theory we suggest the firm use, and the signals indicating that the firm is in fact using the right tool.

TABLE 3-1

Analyzing the Preparation Regimen

Area	Appropriate Tool/Theory	Signals
Strategy making	In uncertain situations, encourage *emergent* forces that allow company to find appropriate market/business model	• Limited fixed-cost infrastructure that encourages experimentation • Demonstrated willingness to adapt to market signals • Business plan that tests rather than assumes
Hiring	Schools of experience in situations company likely to face	• Managers in prior assignments have wrestled with problems that are similar to those the new venture will confront
Funding	Uncertain situations require investors that are patient for growth but impatient for profits	• Values of investors (e.g., need for quick growth) • Relationship between company and investors

Discovering the Right Emergent Strategy

A firm's strategy is much more than a thick binder prepared by a top-flight consulting firm. Its actual strategy is manifest in the products the company introduces, the processes it follows, the services it offers, and the acquisitions it makes. Managers can either dictate strategy from above in a *deliberate* fashion or allow strategy to bubble up from below in an *emergent* fashion.[2]

Encouraging emergent forces can help firms discover the appropriate target market. When attempting new-market disruption, the one thing managers can know for certain is that they do not know how a market will evolve. The use of an emergent strategy process gives them the flexibility to interpret market signals and adjust strategic actions accordingly.

How can you tell when a firm is employing an emergent strategy process? One way is to interview senior management. You also can tell by looking at a firm's actions. Has it built a business that can learn and adapt rather than assume and act? Large, up-front investments tend to force firms to seek large or mass-market customers to support their fixed costs. Smaller up-front investments give firms more flexibility. Initial product architectures that are flexible and configurable allow a company to adapt to emergent forces.

Even in situations that demand large up-front investments, firms can still develop ways to test key assumptions and change their business models accordingly.[3] Investors should have *more* confidence in companies that admit what they don't know and have a plan to turn unknowns into knowns. Watch both what a company's managers say and what they do. Do they say, "This will be a billion-dollar market in five years," or "If *X* and *Y* happen, this will be a billion-dollar market in five years. Here is how we are going to test whether *X* and *Y will* happen"? Are they "staging" investment—investing a chunk of money but not investing another chunk until they make discernable progress?

For example, a great number of companies tried to break into the local telephony market after the 1996 Telecommunications Reform Act. Investors poured billions into business plans promising rapid growth. Almost all of these entrants failed, often quite spectacularly. An unstated—and untested—assumption was that these companies would be able to easily use the local phone company's network elements. Had investors worked backward from the projections in the business plans, they might have pushed companies to test this assumption and learn from the realities of the market. A method of testing that assumption would be to use the VCE theory to assess whether the interfaces that attackers intended to leverage really were specifiable, verifiable, and predictable. Recognizing an absence of modularity, investors would have uncovered faulty assumptions behind many of these business plans and chosen to invest differently.

Similarly, the exponential supply of and growth in demand for bandwidth was a popular untested assumption that became the foundation of many new ventures in the late 1990s. A slowdown in demand left companies with shattered dreams, worthless pro formas, and a lot of nifty office furniture that ended up being recycled on eBay. The widely accepted "certainties" were in fact uncertain.

Alarm bells should ring when an investor sees a well-researched, highly polished business plan and a company showing a great deal of confidence in the numbers in uncertain circumstances. When companies say things like, "Customers just don't get it. They don't know what they want," it indicates that they are clinging to a deliberate strategy in the face of countervailing evidence.

Deliberate strategies have their place. Once the signals from the marketplace are clear, once the winning strategy is clear, companies need to flip to deliberate strategies.[4] *Rigorous* and *disciplined* are both adjectives that should come to mind in these situations. Similarly, deliberate strategies work very well for large established firms pursuing sustaining innovations.

Checking the Schools of Experience

The management team is important. This is not news. One of the first things we are taught to do when evaluating a new venture is to flip to the back of the business plan and look at the founders' résumés. But what should we look for? *The Innovator's Solution* described a theory developed by Professor Morgan McCall called *schools of experience* to help guide hiring decisions. While hiring is an inexact science, the theory holds that managers are *made* more than they are *born*. People have a higher probability of being successful if they have wrestled with a challenge, or attended a school of experience, in the past that is similar to a challenge they will face in the future. The model suggests, in other words, that looking at résumés is good—as long as you are looking for the right things.

Each company faces its own challenges and requires that its management have attended a unique set of schools. The advice given in *The Innovator's Solution* still holds—make a list of challenges you *know* a manager will face and make sure that a manager has come to grips with similar challenges in the past.[5]

What are some of the challenges that we *know* disruptive companies operating in highly uncertain environments must overcome? Managers in disruptive companies should have taken at least some of the following courses in the school of experience:

- Operated in environments with high degrees of uncertainty
- Developed plans to unearth seemingly unattainable knowledge
- Experimented and found unanticipated customers for a product or service
- Placed bets based on theory and intuition, not necessarily detailed data
- Resourcefully solved problems without spending much money
- Built a management team from scratch—a team with skills matched to the task
- Shown experience in fending off certain corporate processes and in harnessing or manipulating others, in order to get the right things done quickly

Managers who have wrestled with at least some of these problems should be well positioned to lead the introduction of disruptive innovations. Of course, as the previous section noted, managers face different challenges when the context calls for deliberate rather than emergent strategies. In other words, managers need to have attended different schools to master these sorts of challenges.

Finding the Right Funding Sources

Every innovator—from the entrepreneur sketching on a Starbucks napkin to an engineer at a multibillion-dollar company—needs funding. Commercializing an innovation takes resources. The investor's values must match the innovation's needs at each point in the growth process.

VCs and Disruption

Our research suggests that the source of money is not a critical determinant of success for recent disruptive ventures. Storage maker EMC was self-funded from its beginning. Cisco only attracted venture funding after it had developed a profitable business model. Bloomberg launched its disruptive terminal using funding from Michael Bloomberg and Merrill Lynch. Charles Schwab couldn't find a venture backer for his discount brokerage model in the 1970s.

Consider the creation of the personal digital assistant (PDA) market. Whose money helped create the market and whose money hurt? Corporate money seems not to have been helpful. Apple's efforts to create the PDA market are a well-studied flop. It invested more than $350 million in creating its Newton product. Apple tried to use a bevy of new technologies to create a miniaturized version of a personal computer, but the product never performed well enough to appeal to customers. Venture capital money didn't help a whole host of would-be disruptors. Kleiner-Perkins backed a start-up called GO Corporation that spent $75 million—and then WENT. All told, these companies and venture capitalists invested more than $1 billion trying to crack the market.

On the other hand, Palm was venture funded. But it was *not* spending its venture capital money that helped Palm find its foothold. Palm spent less than $3 million developing its Palm Pilot. Backed by venture capital funding from Merrill, Pickard, Anderson & Eyre and Sutter Hill Ventures, Palm first created a product called Zoomer PDA, which according to *Fast Company* "did lots of things, most of them badly."* Like the products launched by Apple and GO, the Zoomer

The Innovator's Solution noted that companies introducing disruptive innovations ought to be patient for growth but impatient for profits. Companies will struggle to do this if their investors' values make them want to get very big, very fast.

Many industry watchers simply look for a seal of approval from a "name" investor to separate potential winners from potential losers. Top-shelf venture capitalists (VCs) bring capabilities to the table. They have experience and a wealth of contacts. They see the best deals. Firms that make it through their screens, industry watchers think, must be destined for success.

failed. But Palm had stored up enough cash to develop a second generation. It decided to find out what the handful of Zoomer purchasers actually thought of the product. It stumbled onto a surprising revelation: People did not want a *replacement* for the computer; they wanted a *complement* that was limited but simple to use. Palm came up with a relatively simple device that was limited to applications such as an electronic Rolodex. It forced users to use an intuitive, simple writing style instead of relying on complicated handwriting recognition software. The Palm Pilot seamlessly synchronized with a user's computer, allowing the consumer to have a consistent repository of data. This was a disruptive innovation—simple, cheap, and limited, but positioned squarely on a job that delighted users.

Armed with its winning idea, Palm sought additional VC financing. But purse strings had tightened after the recent explosion of hype surrounding the handheld market. Finally, Palm decided to contact U.S. Robotics, a leading modem manufacturer, who decided to acquire Palm for $44 million in stock.

Palm started the PDA revolution, which created massive amounts of new growth that many believe will ultimately culminate in a product that disrupts leading computer manufacturers. Which funding source should we credit with starting this revolution? Categorizing money as venture capital versus corporate is cutting the world the wrong way. Good money providers are those who help a company spend a little and learn, until it has a viable strategy and business model.

Pat Dillon, "The Next Small Thing: What Does It Take to Change the World? Obsession. Tenacity. And Lots of Mistakes,"** ***Fast Company**, 1 June 1998.**

But investors bring something else: their values. Like all firms, venture capital funds' values change as they grow. Over the 1990s, many VC firms became noticeably more rules-based, data-driven, and insistent on being shown evidence that a market existed before they would consider investment. This made it increasingly difficult for them to follow gut intuition to invest in potentially disruptive opportunities (see "VCs and Disruption").[6] An investor who demands detailed quantitative analysis and market projections might dismiss a company offering a seemingly limited product to an immeasurable market. In the parlance of this chapter, such investors demand execution of deliberate strategy and won't tolerate emergent forces. Furthermore, increasing fund size raised the pressure for investments to grow quickly and hit it big.[7] A natural target for a big hit is a large market. Large, obvious markets tend to be poor homes for disruptive innovations.

Corporate capital can suffer the same drawbacks, especially if corporate growth has stalled. Studies show that once growth stalls, it is very difficult to restart. A large corporation that has stopped growing needs an enormous amount of growth from its new businesses to make the corporation's revenue line look attractive again.[8] But disruptive innovations are like fine wines. They take time to grow and mature.

This suggests looking beyond the *name* of the investor to the *values* of the investor. How do you do this? Assess the needs of the entity from which the money is coming. Is it in a situation in which it can be impatient for profits but patient for growth? Or is its circumstance one that will cause executives to pressure the disruptive venture to get very big, very fast? Is there a match between the values of the investor and the needs of the company seeking funding? Another important factor is the relationship between the investor and the company. Research shows that investors who have arm's-length relationships with the companies they invest in tend to stop the flow of funds when the companies experience some turbulence.[9] We know the disruptive ride is nothing if not turbulent. If you know a company will experience some changes in direction, make sure it has investors who can tolerate them.

One other important component of funding to consider is the amount of money a company raises. People often think bigger is better. They cheer a company that raises a treasure trove of capital. Lots of funding enables the company to make major fixed-cost commitments. It can spend its way to profitability. But think about how that treasure trove affects a firm. First, it must get big fast in order to provide attractive returns to whoever provided all that capital. This would lead it to turn down a small, profitable opportunity that could launch it on a truly disruptive trajectory. Second, too much money allows firms to follow a losing strategy for too long. Businesses that are unprofitable for several years often *never* become profitable. Making profits gives a company the privilege to invest and grow more. Most companies should raise enough initial capital to develop a product to bring to an initial market but not significantly more.[10] The discipline imposed by tight purse strings forces companies to discover customers who *really* value an innovation. In most circumstances, get-big-fast money is only desirable *after* the company identifies the right customer and develops a profitable business model.

Understanding a firm's preparation regimen helps assess whether it will be able to successfully navigate these waters. Encouraging emergent forces helps discover new markets. A management team that has successfully overcome (or at least understands) the temptation to target obvious markets can have an advantage. Investors who encourage experimentation (within reason) and don't push a firm toward a big market help as well.

Even if a firm takes all the right actions and selects the right foothold market, it must avoid a second pitfall: relying on existing value networks to reach customers. Firms have the best chance of creating asymmetries if they reside in or build a freestanding value network, completely separate from their competitors.

Overlapping Value Networks Can Facilitate Incumbent Co-option

All companies reside in a value network, which includes its upstream suppliers; its downstream customers, retailers, and distributors; and its partners and ancillary industry players. Think about a traditional computer manufacturer. It purchases inputs such as microprocessors and monitors. It sells its product to retailers, who sell the product to the end consumer. Postsales support consultants, software companies, and repair shops all provide important ancillary services. The business models firms develop and the rhythms by which their processes address common problems must be compatible across these value network participants.

Strategic Choice: Overlapping Value Network Leads to Co-optable Business Model

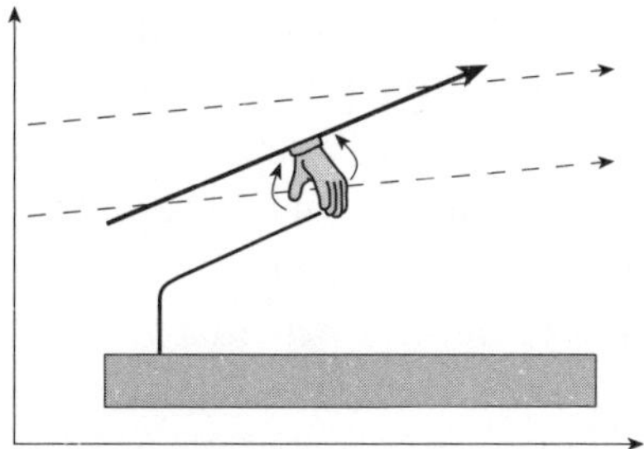

Overlapping value networks with *choke points* can limit the ability to create asymmetries. When there are overlapping suppliers, distribution networks, sales forces, or ancillary providers, firms can face severe pressure to create something that makes sense to the competitor's value network and hence makes sense to the competitor.

How do overlapping value networks lead to symmetrical motivation and skills? Consider a potential low-end disruptor. To have disruptive potential, the entrant needs to develop a different production model or have lower overhead costs to make attractive returns at low price points. But if a would-be disruptor uses the same suppliers as an incumbent, the disruptor will face pressure to adapt to the incumbent's cost structure. Similarly, when an entrant tries to use an existing sales channel or a similar distribution or retail channel to reach an end customer, the entrant's business model *must* conform to the channel's resource-allocation criteria. If not, the sales force and the channel cannot prioritize selling the entrant's product or service.

For instance, emerging wireless firms' decision to make sure subscribers could place calls to and receive calls from existing landline customers forced them to overlap with the wireline telephony value network. A natural choice, to be sure. But when those companies connected their services to the existing value network, they then faced severe pressure to make sure their business model made sense to the value network. They had to work with the local telephone companies to make sure calls would interconnect. They had to develop transfer pricing mechanisms. They had to purchase the same equipment from the same suppliers as their competitors. As we noted in the introduction, this overlapping value network gave incumbents an additional view into the world of wireless, and the business models became consistent with, and therefore attractive to, the incumbent wireline network operators. Complementary value networks often make incumbent response easy. Remember, without asymmetries, co-option replaces retreat as the natural response strategy.

What could emerging wireless companies have done instead? An explicit strategy of disrupting existing telecommunications providers would have required creating a new, *freestanding* value network. Although service providers assumed they had to create a wide-branching network, the vast majority of consumers spend the bulk of their time within an hour of their homes. It would have been theoretically possible to build a very big and potentially disruptive wireless business while minimizing interactions with the wired network.

For example, companies could have offered a wireless service that only connected with other wireless subscribers and did not allow roaming. Perhaps companies could have started by offering service to teenagers who wanted to chat with their friends on opposite sides of the mall or to parents who wanted a way to stay in touch with their children in case of an emergency. These customers wouldn't have been able to pay much. But they would have been delighted with a crummy product. Roaming would not have been critical. They would have tolerated call quality that would have been unacceptable to a businessperson who needed to spend hours on a conference call.

This strategy would have been similar to the one that Alexander Graham Bell employed to disrupt Western Union. He targeted completely new customers who only communicated with other customers who were within a two- to three-mile radius. But in the early days of the wireless industry, firms instead chose to build a complementary system with high degrees of overlap, limiting their ability to develop asymmetries.

Assessing whether a firm's value network will lead to disruption or co-option entails two steps. First make a list of all of the participants in a firm's value network and a list of all of the participants in its competitors' value networks. This identifies points of overlap. Then assess the degree of overlap, particularly how it affects an entrant's cost structure and business model. When overlap is nonexistent or weak, companies have the best chance of harnessing asymmetric motivation and creating asymmetric skills. When firms build complementary value networks with high degrees of overlap, they often make co-option, not up-market retreat, the natural choice for incumbents.[11]

This chapter's final section shows the proactive choices incumbents can make to master the forces of disruption no matter what an entrant decides to do.

Ways for Incumbents to Earn Their Disruptive Black Belts

Seeing an airplane fly does not disprove the law of gravity. Rather, aircraft designers created a vehicle that could counteract gravity by harnessing forces such as lift, drag, and resistance to make flight possible. Would-be pilots attend flight school to learn how to fly and navigate, altering their techniques in the different circumstances in which they find themselves. Similarly, companies can learn strategies that allow them to control disruptive forces. Just as experts in the principles and implementation of the Six Sigma quality movement can earn black belts, firms can earn their own disruptive black belts. Specifically, they can set up separate organizations to launch disruptive counterattacks or develop internal capabilities to create disruptive growth again and again.

Neither of these approaches are "silver bullets." Creating spinouts or trying to develop internal disruptive engines will fail unless incumbents also follow the right preparation regimen and use the right strategy-making process, hire the right managers, and get funding from the right sources.

Strategic Choice: Incumbents That Earn Their Disruptive Black Belts and Learn to Harness Disruption

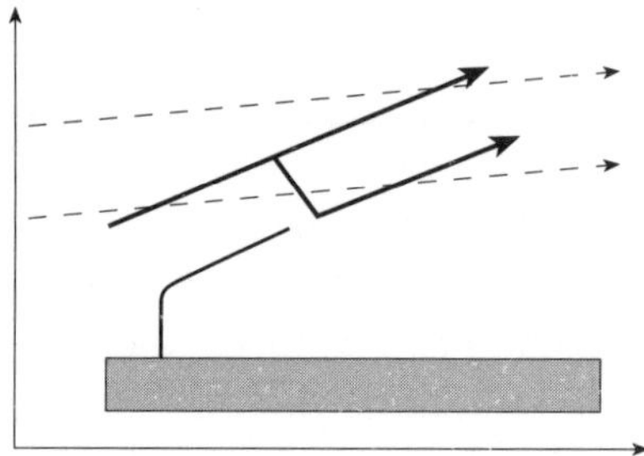

Creating Spinout Organizations to Drive Disruption

Many people think disruption can come only from a completely new entrant. In theory, an incumbent can disrupt itself. It can also create new business ventures that disrupt others. Incumbents that follow the first successful response strategy—the spinout—set up a completely separate business unit free to develop its own skills and define its own metric for success.[12]

Whereas co-option involves trying to fight off a disruptive attacker internally, spinning out a new venture creates an unencumbered external organization to join the fray. Companies with historical success at adapting to disruptive threats followed this prescription. IBM successfully entered the personal computer industry by setting up a separate subsidiary in Florida. Hewlett-Packard ran its disruptive ink-jet business in a separate organization in Vancouver, British Columbia. Intel set up a group in Israel to manage its disruptive Celeron chip. Teradyne set up a separate organization to create its disruptive low-cost semiconductor tester.

Identifying when an incumbent is attempting this strategy is easy. Companies typically issue announcements heralding the arrival of these separate organizations. But seeing a press release is neither necessary nor sufficient. To assess whether or not a spinout strategy will be successful, you need to make sure the incumbent separates the right elements of the spinout, allowing it to create its own values and follow its own preparation regimen. If the company "blesses" its spinout organization with established processes and makes it conform to the overall organization's values, the spinout organization is likely to look very much like the rest of the organization. When a firm subjects a separate organization to the corporate cost structure or project approval process, the new group tends to gravitate toward a sustaining business model. Also, the parent organization often presses for quick growth lest it "pull the plug."[13]

If the incumbent manages these challenges and provides the spinout organization with enough latitude, it can create powerful advantages over other attackers. The incumbent can even stack the deck by supporting the spinout with resources and processes that would actually facilitate its disruptive path.

One very important note: Spinning out an innovation is not a one-size-fits-all, do-it-every-time prescription for managing innovation. In sustaining situations, when incumbents are trying to reach undershot customers, setting up a separate organization denies the new venture the vital fuel of the firm's established skills. It is *only* when a firm has neither the skills to go after an opportunity nor the motivation to develop it internally that a spinout organization makes sense.

Developing the Capability to Build a Disruptive Growth Engine

The second incumbent response strategy involves setting up a repeatable process for managing disruptive innovations internally. Disruption kills incumbents because their processes and values can't simultaneously handle disruptive and sustaining innovations. But a company that can follow the guidelines in *The Innovator's Solution* could set up a set of processes that manages disruptive innovations again and again.

We can't yet point to an example of a firm that actually has become a serial disruptor. A company has its best chance of following this path if it follows the four-pronged approach outlined in chapter 10 of *The Innovator's Solution.* The company should:

1. Start before it needs to
2. Appoint a senior executive to shepherd ideas into the appropriate shaping and resource-allocation process
3. Create a team and a process for shaping ideas
4. Train the troops to identify disruptive ideas

All of these items are important, particularly the second and third. A company bent on becoming a serial disruptor *has* to have a separate process to nurture disruptive growth. The process has to have a set of screens and criteria that determine whether an innovation is in fact disruptive to the mainstream business. This process *must* be managed separately from the sustaining innovation process. A powerful senior manager *must* stand astride the resource-allocation process, channeling disruptive innovations through one process and sustaining innovations through another. Don't expect a company to develop a competency in disruption just because it throws money at the problem. In fact, throwing money at the problem is often exactly the wrong thing to do. Don't expect simply setting up a disruptive department to be the answer, either. Only a robust, repeatable process separated from the mainstream values that stifle the prioritization of disruptive opportunities can create wave after wave of disruptive growth.

Both of these actions—a one-time response of spinning out a counterattack and a long-term response of building internal capabilities to master disruption—can shift the advantage back to the incumbents.

This chapter showed the strategic choices that influence the process of disruption. Entrants can follow a preparation regimen that leads them to target the wrong customers. They can develop complementary business models and capabilities, especially if they reside in an overlapping value network. Both of these events tip the scales in the favor of incumbents. Incumbents who have earned their disruptive black belts can create spinout organizations to parry disruptive threats, or they can develop internal capabilities to harness the forces of disruption again and again.

Key questions to ask when analyzing strategic choices include:

- *Is the company in a situation in which the right strategy needs to emerge? Is the firm giving itself the freedom to encourage emergent forces? Have managers wrestled with problems they are likely to face again? Have they shown the capacity to learn?*
- *Are investor values aligned with company needs? If the investor is a corporation, has growth stalled?*

- *Do value networks overlap? If they do, what are the degrees of overlap? Do they make it impossible to create a business model that has asymmetries?*
- *Is this an appropriate situation for a spinout? Is the company giving the spinout the freedom to do what is necessary?*

Although these chapters presented this process in a linear fashion, picking winners and losers is not so simple. Some issues are codependent. Therefore, theory-based analysts need to constantly watch for signals of change, evaluate competitive battles, and look at strategic choices.

Notes

1. Emergent strategy making is discussed in Clayton M. Christensen and Michael E. Raynor, *The Innovator's Solution: Using Good Theory to Solve the Dilemmas of Growth* (Boston: Harvard Business School Press, 2003), chapter 8. The notion of schools of experience comes from Morgan McCall's excellent book *High Flyers: Developing the Next Generation of Leaders* (Boston: Harvard Business School Press, 1998). The concepts in McCall's book are summarized in Christensen and Raynor, *The Innovator's Solution*, chapter 7, and Clayton M. Christensen and Morgan McCall Jr., "Getting the Right Stuff in the Right Place at the Right Time," Note 601-054 (Boston: Harvard Business School, 2000). The "right" source of funding is discussed in *The Innovator's Solution*, chapter 9.

2. Many of these concepts originated from our friend Robert Burgelman at Stanford University. See Robert A. Burgelman, *Strategy Is Destiny: How Strategy-Making Shapes a Company's Future* (New York: Free Press, 2002).

3. *The Innovator's Solution* notes that discovery-driven planning can be a good tool to use in these situations. See Christensen and Raynor, *The Innovator's Solution*, chapter 8, and Rita Gunther McGrath and Ian C. MacMillan, "Discovery-Driven Planning," *Harvard Business Review*, July–August 1995, 44–56.

4. In reality, no strategy is purely emergent or purely deliberate—every strategy has some elements of both. But companies can differ dramatically on the emphasis they place on both types of forces. A good research effort would unearth indicators that identify when the Rubicon has been crossed and strategies need to tilt more toward the deliberate side of the spectrum.

5. The schools of experience theory always helps to evaluate the likelihood that a manager has the skill to successfully address a problem. Whether you put a fully skilled person in charge or take a risk on someone who has the capacity to learn from past experiences depends on whether immediate performance is paramount.

6. We looked at the funding history for Palm, Bloomberg, Charles Schwab, Southwest Airlines, Polycom, EMC, Dell, Oracle, and Cisco. Cisco, Palm, and Polycom all received venture backing at some point in their existences, but that venture backing was not critical to the creation of their disruptive business models. Some relevant information sources: On Palm: Pat Dillon, "The Next Small Thing: What Does It Take to Change the World? Obsession. Tenacity. And Lots of Mistakes," *Fast Company*, 1 June 1998. On Schwab: Terence P. Pare, "How Schwab Wins Investors," *Fortune*, 1 June 1992. On Polycom: Clayton Christensen, Tara Donovan, and David Sundahl, "Polycom, Inc.: Visualizing Culture," Case 9-601-073 (Boston: Harvard Business School, 2000); Brenda L. Moore, "Video-Conferencing Firm Polycom Looks and Sounds Like a Leader," *Wall Street Journal*, 7 April 1999; Daniel Fisher, "Told Ya So: If One of Your Engineers Says You Should Slash the Price of Your Lead Product by 80%, You'd Better Listen," *Forbes,* 14 December 1998. On EMC: Dana Bottorff, "High Tech Success With Products to Improve and Prolong Computer Life," *New England Business*, 6 July 1987; Aaron Zitner, "Memory, Megabytes and Megabucks: EMC Corp. Finds Building Profits in 'Electronic Filing Cabinets,'" *Boston Globe*, 17 May 1994; Peter Branton, "IBM goes Symbiotic with Biggest Rival," *Network News*, 14 April 1999. On Dell: Charles Boisseau, "The Company That Dell Built—Isn't Finished Yet," *Houston Chronicle*, 25 February 1996. On Cisco: Robert X. Cringley, "Nerds 2.0.1; Serving the Suits," *PBS Online* <http://www.pbs.org/opb/nerds2.0.1/serving_suits/cisco.html> (accessed 15 May 2002).

7. Interestingly, this line of thinking suggests the VC recession in 2000 to 2003 is a great long-term development for the industry. If large funds continue to shrink, VCs can go back to doing what they do best—trusting their instincts and following patterns. If large firms stay big and simply shift to later-stage deals, they will create an opportunity for a new set of funds to pursue smaller deals in a pattern-recognition mode.

8. A report by the Corporate Advisory Board called *Stall Points* documents this finding. It found that once corporations reach a certain size, growth stalls and tends never to return. In this situation, corporations will want immediate payoffs in their investments. For more information, see *Stall Points* (Washington, DC: Corporate Strategy Board, 1998).

9. Harvard Business School professor Myra Hart observed this phenomenon in her research (see Myra Maloney Hart, *Founding Resource Choices: Influences and Effects*, Ph.D. diss., Harvard University, 1995). Interestingly, it could provide another advantage to corporate capital. The corporation is likely to have close relationships with the people in whom it invests. The corporate capital, therefore, might be better able to tolerate growing pains.

10. Some investors think that "staging" investment—doling out portions of an investment at a time—is the same as being patient for growth and impatient for profits. This is not necessarily the case. Just because an investor provides money in chunks does not mean it is encouraging experimentation and learning. In fact, the investor can stage capital in a way that only provides more funding if progress is

made toward a deliberate strategy. This can cause a firm to shut off emergent forces and miss important market signals.

11. Freestanding value networks are particularly important for new-market disruptions. Would-be low-end disruptors can find success using existing additional value network participants as long as that does not constrain their ability to offer lower prices. For example, discount retailers that emerged in the 1960s sourced mainstream products; steel minimills distributed through major steel distributors such as Ryerson and Edgecomb.

12. By definition then, of course, the new organization the incumbent sets up would be classified as an entrant in this market context.

13. It is not always necessary to set up a completely new company where the existing company's only interaction is through equity ownership. A separate business unit will suffice as long as the company does not subject that business unit to the constraints of the mainstream organization.

CHAPTER FOUR

HOW NONMARKET FACTORS AFFECT INNOVATION

What are the true drivers of innovation? How do nonmarket factors affect these drivers? What actions can nonmarket players take to increase the pace of innovation? What actions are likely to impede innovation? Which will have no effect? How can we identify circumstances in which no intervention is necessary, in which proper intervention can be successful, and in which intervention is likely to have little impact?

This chapter continues to use the telecommunications industry to develop the motivation/ability framework introduced in chapter 1 and to explain how to assess the impact of nonmarket intervention on innovation.[1]

One of the stated goals of the U.S. government is to promote the general welfare. As such, the government often tries to identify and address market failures and increase the pace of innovation within an industry. Conventional wisdom suggests that government involvement is bad for innovation. The implicit theory frequently in operation is that more intervention leads to less innovation and less intervention leads to more innovation.

This simplistic view ignores a long list of innovations that would not have been successfully created or exploited without the government's support. For example, the Internet's roots trace back to decades of work by scientists and engineers from the Defense Advanced Research Projects Agency (DARPA) and other government agencies. The government played a vital role in modernizing health care through funding research at the National Institutes of Health.

The government's role in influencing the creation and exploitation of innovation clearly goes well beyond directly financing research and development projects. There is actually an observable and predictable relationship between the natural progress of innovation and the actions governments take to oversee markets.

The Motivation/Ability Framework

In chapter 1, we noted that two factors can be observed in market environments in which innovation thrives: motivation, defined as a pot of gold waiting for the winners, and ability, defined as the capability to obtain resources, craft them into a business model, and offer products and services to customers. Nonmarket factors such as industry standards, unions, cultural norms, the state of technological development, a country's intellectual property infrastructure, and most important, government regulation all can affect would-be innovators' motivation and ability.

The motivation/ability framework is a two-by-two matrix (displayed in figure 4-1) with continuums of motivation and ability along the axes. Government policy (or other nonmarket forces) influences a market's position on each of these axes. Therefore, we can use this grid to assess the impact of any policy, regulation, or other initiative on innovation.

Figure 4-1 suggests four distinct industry situations: the *hotbed*, in which motivation and ability exist in abundance; *looking for a target*, in which innovators lack the ability to either create or exploit an innovation; *looking for the money*, in which innovators lack the motivation to create or exploit an innovation; and the *dilemma*, in which innovators have neither motivation nor ability. Let's discuss each quadrant.

The Hotbed: Innovation Abounds

An abundance of both motivation and ability places an industry in the upper-right-hand quadrant in figure 4-1, termed the hotbed. The hotbed teems with innovation.[2] Both incumbents and new entrants are unfettered, with opportunities for incumbents to pursue profitable sustaining

FIGURE 4-1

The Motivation/Ability Framework

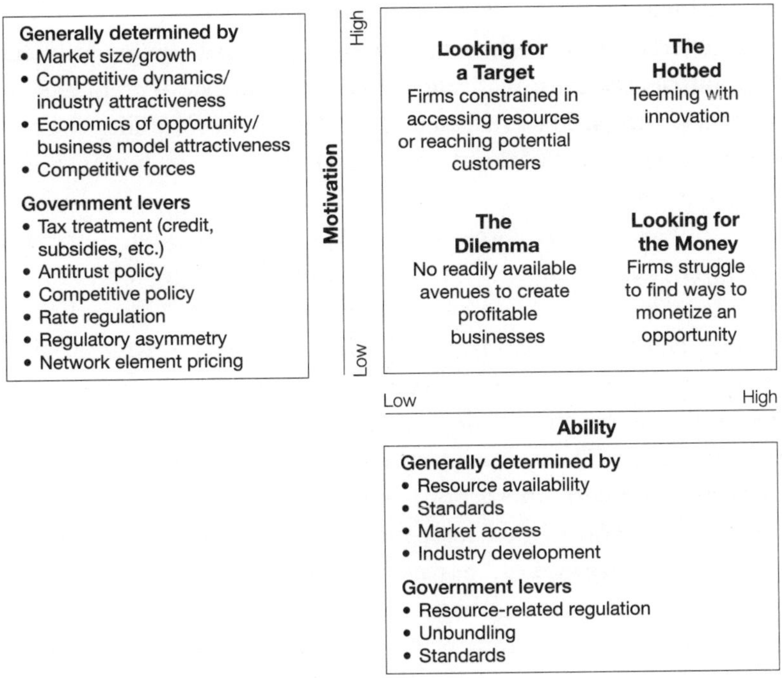

Generally determined by
- Market size/growth
- Competitive dynamics/ industry attractiveness
- Economics of opportunity/ business model attractiveness
- Competitive forces

Government levers
- Tax treatment (credit, subsidies, etc.)
- Antitrust policy
- Competitive policy
- Rate regulation
- Regulatory asymmetry
- Network element pricing

Motivation	Low Ability	High Ability
High	**Looking for a Target** Firms constrained in accessing resources or reaching potential customers	**The Hotbed** Teeming with innovation
Low	**The Dilemma** No readily available avenues to create profitable businesses	**Looking for the Money** Firms struggle to find ways to monetize an opportunity

Ability

Generally determined by
- Resource availability
- Standards
- Market access
- Industry development

Government levers
- Resource-related regulation
- Unbundling
- Standards

innovations and for potentially disruptive entrants to attack leading incumbent firms.[3]

Many technology-related industries historically resided in the hotbed. Consider the rigid disk-drive industry. The industry's rapid growth drove motivation. Almost nothing stopped innovators from developing both sustaining and disruptive innovations. Innovation abounded. This is one reason Christensen selected the industry as the focus for his research that formed the basis of the disruptive innovation theory.

Looking for a Target: Barriers but Possibilities

In looking for a target situation, firms can see a pot of gold at the end of the proverbial rainbow. But something constrains their ability to reach the pot of gold. The government plays a vital role in creating and fixing these situations through its influence over access to necessary inputs and customer groups. Actions that affect things such as public property rights and trade policies further shape entrepreneurs' abilities to move potentially good ideas from the sketchpad to the salesperson.

In an extreme case such as the wireless industry, the government actually controls the critical raw material—the blocks of frequency spectrum required to offer services. Hence, an innovator's ability to come up with and exploit the potential of new offerings is highly dependent on the government.

The troubles of Leap Wireless, a QUALCOMM spinout, illustrate how ability barriers can hamper an innovator's capacity to grow and move upmarket. Leap introduced its Cricket service in October 2000 in Knoxville, Tennessee. Subscribers enjoyed unlimited calling any time of the day from their home markets. They were unable to roam, did not receive detailed breakouts of calls made, and could not use many advanced features (such as wireless Web access) that other wireless providers offered.

Leap had all the hallmarks of a low-end disruption. It targeted overshot customers who tended to spend the majority of their time in and around their homes and did not need buckets of long-distance minutes. Leap's service was low cost, because its prepaid option meant it did not need to conduct costly credit screening or hunt down delinquent accounts. Its limited feature set also helped to minimize its ongoing operational expenses. Leap designed Cricket to profitably serve the lower tiers of the market, which contained incumbents' least profitable subscribers. Everyone from homemakers to students to local proprietors signed up, making Leap one of the nation's fastest-growing wireless companies.[4]

But in April 2003, Leap filed for Chapter 11 bankruptcy protection. At the time of the writing of this book, its stock was trading at less than $0.05. With a promising disruptive business model, what went wrong? Leap launched its product in what it called *island of light markets*—defined as the isolated and self-contained markets, such as Denver, Pittsburgh, or Buffalo, that appear as islands of light as you look at a map of the United States at night. It avoided the top-tier metropolitan markets that represent the bread and butter of incumbent providers. This was an appropriate way to channel asymmetries. But there are only so many island markets. Although Leap could slowly grow in its existing markets, meeting its financial requirements meant it needed to grow faster. Faster growth required obtaining additional swaths of spectrum to offer services in new cities, but the government had already portioned off the available spectrum in key cities. The lack of available spectrum hampered Leap's ability to expand.[5]

Even with limitless resources, entrants and incumbents alike cannot create growth from innovative offerings if government policies prevent them from reaching customers. Some of the tools the government has in its arsenal, such as licensing and minimum quality standards requirements, can affect the capacity of firms to offer services by controlling their ability to access groups of customers, especially the less demanding customers who are the hallmarks of disruption.

Although there is a role for oversight because unrestricted and unchecked access to customers can lead to consumer harm, allowing innovators to reach customers can also create significant consumer welfare. For example, the government denied AT&T the right to offer voice mail services in the late 1970s. Regulators considered voice mail an "information service." AT&T could not legally sell information services. The Modified Final Judgment (MFJ) court finally relented in 1988 and allowed the local telephone companies to offer the service. Massachusetts Institute of Technology professor Jerry Hausman estimated that this regulatory delay destroyed about $1 billion of consumer welfare.[6] The voice mail case shows how efforts to protect consumers actually denied them the opportunity to benefit by getting access to cheaper, more convenient technology that greatly enhanced their lives.

Because these sorts of barriers often protect the lowest tier in the market, they exacerbate tendencies to cram potentially disruptive innovations into the highest market tier. Relatively low-performing disruptive innovations tend to hit minimum quality standard screens. An innovator prevented from selling her potentially disruptive product to a group of customers may be forced to turn to deep-pocketed incumbents, the only channel capable of reaching those customers. Chapter 2 showed how incumbents tend to cram disruptive innovations, snuffing out the inherent disruptive energy. This effect is particularly pernicious in the health care industry, because stringent Food and Drug Administration requirements mandate that products meet the needs of the most demanding customers. This tends to encourage entrepreneurs in regulated industries to mutate potentially disruptive ideas into one-size-fits-all solutions.

Hope is not lost in the looking for a target quadrant. Given enough motivation, creative engineers and strategists will find ways to circumvent or overcome even the most intimidating and discouraging constraints on ability. For example, innovators constantly break theoretical limits that appear to place a ceiling on the improvement potential of optical lithography, an essential part of the microchip manufacturing process. Scientists regularly invent new and previously unimagined ways to overcome what experts perceive to be well-established and fundamental laws of nature.[7] By improving optical lithography's capacity to etch more circuits on a single chip, processor speed can increase according to Moore's Law (discussed in chapter 7).

Looking for the Money: Innovation Is Possible but Not Likely

When firms have the ability to create and offer services to customers but no motivation to do so, they are in the looking for the money quadrant in the lower right-hand portion of the matrix. Here, an opportunity to innovate exists but firms struggle to find ways to monetize it. Without the necessary fuel for the entrepreneurial spirit, successful innovation is unlikely.

Factors such as a market's size and prospects for growth, the general industry attractiveness, the attractiveness of specific business models within that industry, and the level of competition determine motivation.

The government can shape the motivation of both new entrants and incumbents through a variety of legislative and regulatory means. Examples of these motivation tools in the telecommunications industry include rate regulation, interconnection or component pricing, tax policy, antitrust policy, and competitive policy. These mechanisms set the boundaries within which innovators can design business models.

The government can also design policy and regulatory actions to apply to all industry participants or to just a subset. Government has often relied on forms of asymmetric action—policy or regulation that creates motivation for one group while simultaneously diminishing it for another—to increase competition in regulated markets or markets dominated by a few large players.

The government typically targets mechanisms that affect motivation at incumbent firms. For instance, in the energy and telecommunications industries, price caps that dictate the highest price utilities or telephone companies can charge create powerful incentives for companies to improve operational efficiency through cost-cutting and cost-saving innovations. The more a firm can improve operating efficiencies, the more profit it can retain. This is in contrast to return-rate regulation that limits profits by assigning a fixed rate of return. Some critics argue that this reduces the motivation of incumbents to invest in new business opportunities because government mandate limits their potential upside. Proponents argue that return-rate regulation promotes innovation by providing coverage for expensive and risky technological upgrades.

Historical regulation to keep residential telephony prices low to promote universal access at least partially explains why it has been difficult to encourage competition in the local telephony market. The government's regulatory agenda means that new entrants often have little motivation to offer local telephony services.

Having the ability to obtain and transform resources into products desired by customers occasionally exposes formerly unforeseen and even counterintuitive ways to earn profits. This notion continues to prove itself on the Internet. The ability to "hang a shingle" out on the World Wide Web for anyone to find has unexpectedly spawned the creation of new business models. Who knew that an online marketplace where people traded Beanie Babies and Pez dispensers—eBay—would turn into one of the Internet's most profitable businesses?

The Dilemma: Lack of Innovation

In the quadrant opposite the hotbed lies the dilemma—in which firms have neither the motivation nor the ability to innovate. The framework predicts that innovators will be unable to create and exploit innovations in these contexts.

Summing Up Motivation/Ability

Our framework suggests that the creation and exploitation of innovation flourishes in the hotbed; that innovation can still happen in the looking for a target quadrant because motivation can trump ability; that innovation is less likely in the looking for the money quadrant; and that the dilemma is a death trap for innovation.

Chapter 1 suggested looking to see if nonmarket players are taking actions to address barriers to innovation. For example, government action to promote ability led to successful innovation in the CPE and long-distance markets. The government's hands-off policy helped to spur the growth of the Internet.[8]

Understanding how to assess the impact of government policies requires digging a bit deeper. The next section analyzes the 1996 Telecommunications Reform Act and shows how sometimes seemingly appropriate actions fail to solve underlying problems.

The Government and Its Efforts to Change the Game: The 1996 Telecommunications Reform Act Dissected

The landmark Telecommunications Act of 1996 (Telecommunications Reform Act or TRA) seemed to be an effort by the government to address the missing inputs to innovation and push the industry to the hotbed.[9] Theory would suggest success. Indeed, competition grew. Cable companies emerged as a threat. Prices in some local areas fell. Yet many consider the act a failure. Detailed analysis of the act shows how the government, in its efforts to do good, sometimes can do harm. It suggests three lessons: (1) creating motivation can be both difficult and dangerous; (2) creating ability can be more difficult than anticipated; and (3) trying to do both at once can be disastrous.

The analysis below uses acronyms and terms that may be unfamiliar to many readers. Figure 4-2 describes the alphabet soup of acronyms that populate the U.S. telecommunications industry.

FIGURE 4-2

Glossary

CLEC	Competitive Local Exchange Carrier	New entrants
IXC	Interexchange Carrier	Long-distance companies (MCI, AT&T, Sprint)
PSTN	Public Switched Telephone Network	The phone network
RBOC	Regional Bell Operating Company	Local phone companies (Verizon, SBC, etc.)
TELRIC	Total Element Long-Run Incremental Cost	Pricing mechanism determined by FCC for UNEs
TRA	Telecommunications Reform Act	1996 act deregulating industry
UNE	Unbundled Network Element	Individual piece of RBOC network that CLECs could license at TELRIC

Setting the Stage

The act had dramatic effects on the U.S. economy. Acknowledging that the existing regulatory structure was outdated, the drafters sought to encourage growth and innovation with competition and free-market mechanisms. Many laud the policy initiative for its sweeping pro-competitive reforms and the waves of innovation that followed in its wake. Naysayers point to the bursting of the so-called dot-com and telecommunications bubbles, the failure of the competitive local exchange carriers (the CLECs, discussed briefly in the introduction), and the evaporation of a gargantuan amount of invested capital as evidence of the act's shortcomings. The reality is that it did indeed seed certain types of profitable innovation, even while it failed on other dimensions. Here's why.

In 1995, the telecommunications world was a very different place. There was no competition in all but the most densely populated residential local exchanges. The Regional Bell Operating Companies (RBOCs) lorded over a protected monopoly on telephone services. Regulation prevented cable companies and electric utilities from offering telephony services (which they couldn't do anyway given the current state of technology). Wireless technologies were just beginning to mature. The Internet was just beginning to emerge.

How would the motivation/ability matrix classify potential competitors for residential local exchange telephony services in 1995?[10] The RBOCs were the only firms with the ability to innovate in the local telephony market because of their proprietary access. However, they largely viewed residential telephony as a cash cow to be milked. Their primary motivation was to find a way into the then-immensely profitable long-distance oligopoly—a market that existing regulation prevented them from entering. The long-distance companies (known in the industry as interexchange carriers, or IXCs) were motivated to reach customers in the local exchange but lacked ability.

All other potential competitors had relatively low levels of both motivation and ability. Cable companies probably had the most motivation of all potential entrants. Industry analysts viewed telephone service as a logical growth vehicle. Cable companies believed the ability to provide customers with an integrated voice, video, and data service offering on a single bill would be the ticket to growth. It would create additional revenue streams and increase customer loyalty. And cable companies already had a set of wires running into almost every home in the United States.

In enacting the 1996 act, the government hoped to spur competition by breaking down barriers through further deregulation. The act's stated goal was "to promote competition and reduce regulation in order to secure lower prices and higher quality services for American telecommunications consumers and encourage the rapid deployment of new telecommunications technologies."[11] The government seemed implicitly to equate intensified competition with innovation in a causal theory. Interestingly, AT&T and Bell Labs had been exquisitely innovative for decades, creating by far the world's best telecommunications system, even in the absence of competition.

Regardless, the act represented a major overhaul of the telecommunications regulatory regime fashioned by the 1934 Communications Act. The 1996 act, while intended to be deregulatory in nature, reregulated the industry by introducing a new legal framework. While the 100-plus pages of the act included a plethora of specific initiatives, a few of the key provisions included:

- Granting permission for cable TV companies, electric utilities, broadcasters, IXCs, and competitive access providers (CAPs) to sell local services.
- Forcing RBOCs to resell pieces of their local networks (called unbundled network elements, or UNEs) to competitors. The FCC later determined pricing for UNEs based on a provision called total element long-run incremental cost (TELRIC). As interpreted by the FCC, TELRIC prices estimated the forward-looking cost of a best-in-class network technology, including depreciation charges.[12]
- Allowing RBOCs to enter the long-distance market provided they opened their local exchange market to competition. RBOCs could petition the FCC for permission to offer long-distance service once the RBOC proved that its local market was open for competition by satisfying a fourteen-point checklist (known as section 271).
- Granting permission for RBOCs to offer advanced services such as high-speed access to the Internet through a separate subsidiary.

The government tried to push all the players in the industry toward the hotbed quadrant by simultaneously enhancing ability and motivation (figure 4-3 shows what the government attempted to do and what resulted). Officials hoped to increase *ability* for CLECs, IXCs, and cable companies by enforcing local-loop unbundling requirements and removing restrictions that prevented cable companies from offering residential telephony services. Regulators (the FCC) later attempted to create *motivation* for new entrants by introducing the TELRIC pricing mechanism for the purchase of local network elements. Exclusion from common carrier regulations motivated cable companies to enter the fray. Cable companies could use their own networks without worrying about whether they too would need to offer pieces of that network to entrants. Had these prescriptions been enough to create both motivation and ability, IXCs, CLECs, and cable companies would have ended up in the hotbed with the resultant boom in innovation.

The act produced some unanticipated results. Part of the problem was that there was no unanimity about the desired outcome.[13] Some

FIGURE 4-3

Intent and Result of 1996 Telecommunications Reform Act

INTENT

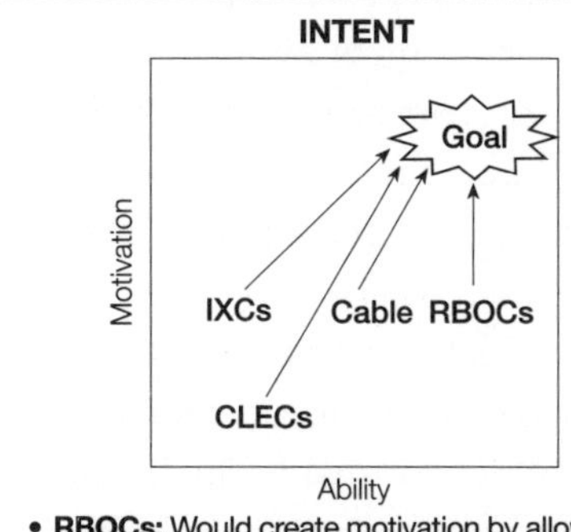

- **RBOCs:** Would create motivation by allowing them to enter long-distance market in return for opening local market to competition
- **CLECs/IXCs:** Would create ability by unbundling local loop (UNEs); would create motivation for new entrants by using TELRIC pricing mechanism
- **Cable:** Would be permitted to offer basic telephony services and not be regulated as a common carrier

RESULT

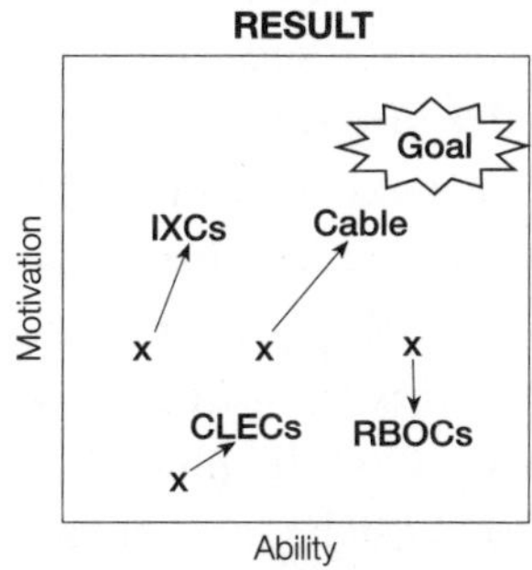

- **RBOCs:** Motivation decreased because of TELRIC optionality, decline in long-distance prices
- **CLECs/IXCs:** Legal ability did not lead to technological ability due to legacy interdependencies
- **All:** Motivation less than desired because of fundamental economics; companies tried to game the system

officials hoped to encourage two to three new facilities-based competitors in the local exchange market. Their model was the long-distance market, where MCI and Sprint aggressively built their own long-haul networks and used the RBOCs' networks only for local transmissions. Other officials wanted to allow new entrants to be able to get to scale as quickly as possible. They hoped entrants would aggressively purchase UNEs to quickly piece together a service offering.

The market expected legislated ability to translate quickly into profitable opportunities. Hundreds of companies entered the market. Exuberant capital markets poured money into CLECs, forcing companies to step up their growth rates. Venture capitalists poured close to $10 billion into start-ups between 1998 and 2000. Publicly traded CLECs had a market capitalization of close to $90 billion at the end of 1999.

But CLECs quickly ran into problems. Most CLECs invested little of their capital in building their own networks. They attempted to piece together a network largely of leased components. CLECs found it difficult to plug into the RBOCs' networks. RBOCs did not cooperate as anticipated. Policymakers hoped to coerce the RBOCs by dangling the carrot of entry into the long-distance market. But increasing competition (largely from the growth of wireless) drove down long-distance prices. The long-distance carrot ended up looking much less appetizing.

Policymakers anticipated some RBOC resistance. But the technological obstacles were larger than expected. Working through the challenges related to integrating with existing networks proved to be extremely complicated. Most investors did not understand the time required to establish and scale a business given the technological complexity involved in provisioning and delivering service.

Profits were tough to come by. As the technology bubble burst, investors pulled the plug. About 50 CLECs went bankrupt between 1996 and 2003.[14] The market capitalization of publicly traded CLECs plunged to about $4 billion at the end of 2001. Although entrants survived in some areas, the act did not lead to the immediate increase in competition and flood of game-changing innovation that the government had intended.

The act's results stymied and frustrated many policymakers. However, the results illustrate three essential principles that must be understood when analyzing the likely impact of any government policy:

1. Creating the right kind of motivation is hard because noise can be mistaken for signal, and root causes can be difficult to address.
2. Creating ability is hard because granting legal ability does not necessarily create operational or technical ability.
3. Legislating out of the dilemma is extremely difficult.

Principle 1: Creating the Right Kind of Motivation Is Hard

When the government corrects a real market failure, it can create motivation capable of stimulating tremendous innovation. However, creating the right kind of motivation is extremely difficult. First, the government often has difficulty separating out true signals that a market isn't functioning optimally from noise that might suggest that the malfunctioning market has a different root cause. Second, even if the government can identify the root cause, it may find that cause difficult to address. As a result, governments have a rich history of creating temporary opportunities that result in a great deal of hype but little real change. Analysts who observe a government meddling with motivation need to be wary of the law of unintended consequences; many legislated advantages can be temporary and can encourage creative entrepreneurs to game the system and exploit the opportunity provided by the government.

Government action to try to increase the motivation of potential competitors is typically a reaction to situations in which the government observes a paucity of new-market entrants. Policymakers often interpret the characteristics that define these situations as a signal of a competitive imbalance that needs to be rectified. When is lack of competition a real signal of a poorly functioning market that limits the natural process of innovation? One situation would be when an existing regulatory scheme or some form of externality creates a market imbalance. When is the apparent signal really noise? When the root cause is either basic industry economics or an interdependent system that stymies specialist new entrants (discussed in principle 2 below). Distorted industry economics clearly constrained innovation in the local residential telephony market. The numerous pricing distortions in the market created by decades of regulation made investment in the local market appear unattractive to most new entrants.

Actions to correct a signal and correct an imbalance can create real motivation, whereas reactions to noise or failure to address the root problem often create artificial motivation. This distinction is important. People usually try to influence behavior by giving some sort of financial incentive, such as subsidies or grants. Our belief is that real motivation comes from a fair and equal chance to earn returns in a competitive marketplace.

Artificial motivation is akin to the government's putting gold on the table to attempt to encourage certain behaviors on the part of managers. Like the parents who find that their teenager excels in doing exactly what they say not to do, artificial forms of motivation tend to encourage behavior that policymakers often are unable to anticipate. A market-based system is very good at directing resources to exploit an opportunity. Firms often come up with novel ways to circumvent regulations and grab the gold without addressing what the government had hoped to correct.

Untangling years of cross-subsidization schemes and regulatory rate caps that discouraged investment was a daunting challenge.[15] Had the government truly tried to address the fundamental problem in the local telephony market, it would likely have resulted in unpalatable consumer price increases. By creating artificial motivation via TELRIC, the government instead encouraged CLECs to develop nonsustainable business models. There are numerous additional telecommunications examples of firms developing tariff arbitrage or cream-skimming business models that take advantage of regulatory regimes in unforeseen ways.[16]

Can efforts to create artificial motivation ever have a chance of being successful? Only when the government and investors have the stamina to patiently stick with a potential business model or innovation as competitive dynamics within a market change (such as in the case of Airbus, discussed in chapter 6). Patience can give artificial motivation a chance to morph into actual motivation. But investors proved unwilling to be patient for the CLECs to find a viable strategy. The government might in fact be the only entity capable of sustaining such a long-term view.[17] Many contend that MCI's emergence as a viable long-distance competitor to AT&T required many years of implicit protection by asymmetric regulation.

Principle 2: Legal Ability Does Not Always Create Technological or Operational Ability

The government often defaults to providing ability to new entrants because it appears to be the easiest policy prescription. Giving new entrants more scarce resources or forcing incumbents to unbundle and resell portions of their fixed infrastructure to new entrants appear to be relatively straightforward exercises. The government can be very successful when a legal barrier is the only thing stopping a firm from entering the market. It is much less successful when technological or operational barriers create the real bottleneck.

How can we tell that legal barriers create the bottleneck? When new entrants, often with market-ready innovations, are jostling at the gates of a market, trying to find ways to get in. In these situations, we can have confidence that the government can promote innovation and competition by identifying and releasing interfaces ready to become modular.

The wireless industry has traditionally been a relative hotbed of competition because in the mid-1990s the government decided to allow more companies to offer cellular services. This led to a dynamic, competitive marketplace. In the wireless data market, the government has allowed firms to operate in so-called unlicensed spectrum (in the 2.5- to 5-gigahertz frequency band) without obtaining approval. This decision created a frenzy of competitive activity by companies attempting to commercialize innovations successfully.

Had CLECs been easily able to interconnect their technology with unbundled elements, the 1996 act might have been similarly successful. However, just because a company has the *legal* ability to do something does not necessarily mean it has the *technological* or *operational* ability. Forcing modularity within the RBOCs' central offices proved to be a complex and lengthy process. The CLECs had to integrate their technology with proprietary network architectures riddled with interdependencies left over from decades of monopoly control.

Theoretically, the local network should have consisted of highly mature technologies that should have been easily modularized. But the absence of competition—the primary driver of the shift from interdependence to modularity—allowed the RBOCs to maintain highly interdependent systems.[18] Without competition, an incumbent has much less incentive to standardize any aspect of its operations. This shield that protects monopolists from competitive forces allows it to continue to squeeze out additional profitability by solving problems in an interdependent way. Stories abound of Bell System workers coming up with novel solutions because they intimately understood the subtle intricacies of the telecommunications network. Their implicit knowledge of the complex interactions within the system enabled them to solve problems that outsiders often could not.

Legacy interdependencies ranged from gaining the presumably simple knowledge of the quality and condition of deployed assets to the more complex job of integrating with existing operation support systems. RBOCs' understandable resistance aside, legacy technological interdependence combined with overambitious build-out plans based on very uncertain demand by firms flush with "bubble levels" of funding contributed to spectacular business flameouts.

This was the critical difference between MCI and the CLECs. MCI built its own network and connected to the rest of the network at an interface point that was specifiable, measurable, and predictable (the three tests of modularity discussed in chapter 1). This is because AT&T had essentially created two autonomous business units (local and long distance) that interconnected in a clean, modular way before MCI entered. The CLECs largely did not build their own networks and connected with incumbent networks in ways that previously had not been defined and remained rife with legacy interdependencies.

Where was the act more successful in fostering competition? In recent years, cable companies have emerged as a viable competitive threat. (We will discuss cable companies in depth in chapter 10.)[19] Here, government actions to promote ability led to innovation.

The technological interdependence that foiled the CLECs was not present for cable companies. They already had an infrastructure that passed just about every home. They had little need to lease elements of the RBOCs' networks. Not constrained by the structural and economic impediments that came with using the existing telephone network, cable operators were free to make the necessary network upgrades to offer telephony services. Although the costs of cable network upgrades were significant, cable companies' ownership of the entire infrastructure afforded them flexibility in when and where they deployed the technology.

Furthermore, as CLEC after CLEC fell, the government began to view cable companies as the most viable means of creating competition in the local market. FCC chairman Michael Powell indicated on multiple occasions his intention to take action to avoid the one-wire scenario. In a speech in 2001, Powell said government should "work to keep multiple platforms and routes to the home open and viable."[20] Implicit in this statement was the belief that cable, as one of the other wires into the home, could play a major role in the future of the industry for both broadband and telephony services.

Principle 3: Getting Out of the Dilemma Is Hard and Takes Time

The government faces the most difficult situation in a dilemma quadrant. Here we should expect government actions to take a long time and not overstate their likely success.

The government can help companies escape the dilemma by focusing on basic scientific research. The costs to pursue these types of opportunities are typically large, up front, and sustained over long periods. The outcome is uncertain. The technologies under investigation are either unproven or nonexistent. Researchers have only the hope of discovery or the desire to learn as their reward. Markets can fail to divert resources to such exploratory endeavors, leaving it up to institutions such as the government to determine whether an opportunity is worth pursuing. Certainly, no one would disagree that there are numerous historic examples of important innovations funded by this type of investment.

In existing markets with stagnant competition, we should be cautious when we see the government rapidly trying to create a competitive environment. There is a natural tendency to try to do everything and encourage ability and motivation simultaneously. These efforts almost always fail, especially given the compromising nature of the policymaking process. Horse-trading efforts to get bills passed lead to watered-down attempts to do everything that often end with serious unforeseen consequences, encouraging pernicious gaming on the part of incumbents or the creation of flimsy business models that take advantage of short-term opportunities created by regulatory change on the part of new entrants.

In this situation, there are two different approaches the government can take that we can expect to be more successful than a "big bang" approach:

1. *Concentrating on one of the two underlying problems.* This approach entails creating an environment where entrepreneurs can figure out a way around the second problem. Although motivation is a more powerful force, its unpredictable nature can make it a dangerous policy tool. Hence, the model predicts that we should have more confidence when the government starts

with policies that promote ability, especially if it can identify a class of players that might be motivated to enter if they just had the ability.

2. ***Creating policies that accelerate the process of disruption.*** When there are simply too many barriers to innovation, barriers that are so strongly entrenched that they cannot be changed, the best strategy is to encourage disruption. Disruption allows firms to spring out of adjacent markets and force radical changes to the market, where change seems to happen at such a glacial pace. A do-nothing approach can actually facilitate disruption by forcing companies to seek other avenues to fund the development of their products. With a hands-off approach, frustrated entrepreneurs will exit the market in which they have neither ability nor motivation and seek out a different market or context where consumers are delighted to embrace what to mainstream customers appear to be relatively poor-performing products.

Summary of TRA Analysis

The act succeeded in some regards. RBOCs began innovating more, seeking to enter new markets. Competition in some local markets increased. The Internet boomed. Cable companies emerged as potential players in the local telephony market. But some factors limited success. Artificial motivation led to unanticipated consequences. Entrants encountered legacy interdependencies; legal ability did not immediately turn into technological ability. What approach would the model suggest?

Perhaps the act should have been less sweeping, focusing on a single class of players rather than trying to influence the behavior of different classes of potential entrants. Perhaps the government should have forced RBOCs to separate into two organizations—one for transport and one for marketing and services (a concept known as *structural separation*). Because the government could have predicted the technological difficulties the CLECs ran into, it could have tempered its enthusiasm about the opportunity it claimed to be creating. With a slower, more rational approach, the government might have introduced greater amounts of sustainable competition in the local market.

Using the Motivation/Ability Framework

Chapter 1 presented a three-stage process for using the motivation/ability framework:

1. Map existing motivation and ability to see if the context is favorable to innovation; if it is not, identify the primary barriers to innovation.
2. Determine if any nonmarket player is taking action to influence motivation and ability.
3. Determine if the action addresses the primary barriers to innovation. If it does, expect the action to promote innovation. Table 4-1 summarizes how this chapter shows how to distinguish circumstances when proper intervention can be successful

TABLE 4-1

When Intervention Can Be Successful

Situation	Identifiers	What Could Happen	Signals
Nonmarket forces *can* improve context	• Companies with demonstrated technologies that do not enter a market due to nonmarket barriers: – Held back due to legal barriers to ability – Not motivated because of distorted economics	Correct action to increase ability or motivation	• Nonmarket innovation targeted at legal barrier or intended to fix source of distortion • Successful new entrants
		Actions that do not address distorted economics and create artificial motivation	• Entrants with business models designed to exploit regulation (gaming)
Nonmarket forces *cannot easily* improve context	• New entrants lack technological ability to enter a market (interdependence required)	Actions to increase ability fail	• Entrants struggle despite increase in legal ability
	• Industry lacks both motivation and ability	Push for "big bang" struggles	• Final version of legislation and regulation features a little bit of everything and a lot of nothing • Entrants with business models designed to exploit regulation (gaming)

(legal barriers to innovation or distorted economics) from circumstances when intervention is likely to have little impact (technological barriers and the dilemma).

Nonmarket forces predictably influence the forces of innovation. They affect either the motivation or ability of innovators to develop and exploit novel products and services. Actions that increase ability or motivation tend to increase innovation; actions that put up barriers to ability or motivation tend to decrease innovation.

The motivation/ability framework can be a helpful tool to understand barriers to innovation as well as how intervention can affect innovation. When barriers to innovation are market imbalances (dampening motivation) or legal barriers (limiting ability), government intervention can push an industry toward the hotbed. In situations in which the barriers to innovation are technological barriers or fundamentally poor economics, government intervention is less likely to push an industry toward the hotbed.

Remember, just because an industry is in a situation in which nonmarket actions can *successfully promote innovation, it is no guarantee that nonmarket forces* will *successfully promote innovation. Nonmarket players must take the correct action to target the barrier. Sometimes that action is politically unpalatable (for example, when fixing motivation requires short-term price increases). Unfortunately, in these situations the politically palatable action can exacerbate the problem.*

And the more severe the problem, the less likely it is that a single action can fix it. Government's best bet in the policymaker's dilemma is to encourage disruptive innovators to form a new value network that can ultimately change the seemingly unchangeable industry.

Notes

1. We define government intervention to encompass both legislation and regulation. There are numerous seminal economic works to understand how governments affect industries (specifically telecommunications). For example, see Stephen G. Breyer, *Regulation and Its Reform* (Cambridge, MA: Harvard Univer-

sity Press, 1982); Gerald W. Brock, *Telecommunication Policy for the Information Age: From Monopoly to Competition* (Cambridge, MA: Harvard University Press, 1994); Alfred E. Kahn, *The Economics of Regulation: Principles and Institutions* (New York: Wiley, 1970); John R. McNamara, *The Economics of Innovation in the Telecommunications Industry* (New York: Quorum Books, 1991); Richard A. Posner, *Natural Monopoly and Its Regulation* (Washington, DC: Cato Institute, 1999); William W. Sharkey, *The Theory of Natural Monopoly* (New York: Cambridge University Press, 1982); and John T. Wenders, *The Economics of Telecommunications: Theory and Policy* (Cambridge, MA: Ballinger, 1987). All of these works are very useful. However, our observation is that few researchers attempted to tackle the specific question of how government intervention affects innovation.

2. It is important to note that whereas hotbed environments provide innovators the greatest chance of creating sustainable, profitable enterprises, not all innovations will necessarily be successful.

3. In practice, of course, the ability to innovate is not always equally shared among all existing and potential competitors. In many markets, incumbents have the upper hand and have established barriers that prevent the successful entry of new competitors.

4. Interestingly, surveys showed that customers choosing Leap eschewed traditional landline service. Leap, press release, 24 June 24 2002, <http://www.leapwireless.com/press/content/2002/062402.html> (accessed 10 July 2002).

5. Of course, Leap could have purchased spectrum from existing players. Leap's debt-laden capital structure coupled with investor sentiment reduced its ability to obtain additional resources to purchase more wireless spectrum. When Qualcomm spun out Leap, it left Leap with a substantial amount of debt related to its acquisition of licenses to operate a CDMA-based business.

6. Jerry A. Hausman, "Valuing the Effect of Regulation in New Services in Telecommunications," *Brookings Papers on Economic Activity: Microeconomics* (Washington, DC: The Brookings Institution, 1997).

7. Almost like clockwork, newspaper articles appear roughly every five years with quotes from industry experts claiming that optical lithography technology is about to reach its theoretical limit. The key measure of improvement in the industry is the width of the transistor on a microchip. This width is measured in microns, or one one-hundredth the width of a human hair. Using tightly focused wavelengths of light, optical lithography etches circuit patterns onto a silicon chip through a sophisticated template. Making smaller components capable of running faster microchips means focusing light with greater precision, which is expensive and difficult. In the 1970s, experts thought the theoretical limit to be about one micron. In the mid-1980s, scientists believed the limit was 0.3 to 0.4 micron. By the 1990s, experts redefined the limit to 0.18 micron. As we entered the twenty-first century, industry insiders were once again hesitating to make predictions. See Ronald Rosenberg, "Chip Makers Turning Toward X-ray: With the 4-Megabit Semiconductor, Optical Lithography Has Reached Limit," *Boston Globe*, 8 Au-

gust 1988; Dwight B. Davis, "Technologies Ride a Fast Track into the 1990s," *Electronic Business Buyer*, 11 December 1989; David A. Markle, "Lithography: The Road Ahead," *Solid State Technology*, 1 February 1999.

A quote from an article in *Electronics Weekly* sums up this phenomenon: "Anyone involved in the chip making business can remember someone at sometime saying that the economics of commercial chip making would be stood on their head in x-years time when optical lithography would reach its limits. Everyone recalls those statements with a smile. For optical lithography has surprised everyone by having more stretch in it than anyone intended." See "JESSI Looks to Optical Lithography to Take a 0.18 Micron Curtain Call," *Electronics Weekly*, 16 November 1994.

8. For an excellent study of this, see Jason Oxman, "The FCC and the Unregulation of the Internet," working paper 31, Federal Communications Commission Office of Plans and Policy, Washington, DC, July 1999.

9. We owe a debt to a number of remarkably cogent analyses of the 1996 Telecommunications Reform Act. For example, see Robert W. Crandall, "An Assessment of the Competitive Local Exchange Carriers Five Years After the Passage of the Telecommunications Act," *Criterion Economics*, 27 June 2001; Robert W. Crandall, "Reply Declaration in the Matter of Implementation of the Local Competition Provisions in the Telecommunications Act of 1996, CC Dkt. No. 96-98," 30 April 2001, <http://www.criterioneconomics.com/docs/crandall_final.pdf> (accessed 25 February 2004); Robert W. Crandall, "Are We Deregulating Telephone Services? Think Again," Brookings Institute Policy Brief 13, March 1997, <http//www.brook.edu/comm./PolicyBriefs/pb013/pb13.htm> (accessed 25 February 2004) March 1997; James K. Glassman and William H. Lehr, "The Economics of the Tauzin-Dingell Bill: Theory and Evidence," working paper 128, Massachusetts Institute of Technology, Cambridge, MA, 11 June 2001, <http://ebusiness.mit.edu/research/papers/128%20Lehr,%20Tauzin-Dingell.pdf>; Thomas Hazlett, "Economic and Political Consequences of the 1996 Telecommunications Act," working paper 99-8, AEI-Brookings Joint Center for Regulatory Studies, Washington, DC, September 1999; Reed Hundt, *You Say You Want a Revolution? A Story of Information-Age Politics* (New Haven, CT: Yale University Press, 2000); Alfred E. Kahn, Timothy J. Tardiff, and Dennis L. Weisman, "The Telecommunications Act at Three Years: An Economic Evaluation of Its Implementation by the Federal Communications Commission," *Information Economics and Policy* 11 (1999): 319–365.

10. This analysis focuses on the provision of local telephony to the residential market because that provides the clearest examples of both success and failure. Analyzing the business market leads to similar, albeit less stark, conclusions.

11. *Telecommunications Act of 1996*, Public Law 104-104 [S. 652] (8 February 1996).

12. The RBOCs argued that TELRIC was equivalent to the government stealing their assets because they had to lease them at an unreasonably low price.

The Supreme Court disagreed and upheld the provision of TELRIC, although it said its job was not to determine the best pricing scheme, just the legality of the TELRIC provision. As of the writing of this book, courts were still debating precisely what regulators could and could not legally do.

13. An anecdote that then-FCC chair Reed Hundt recounted in his autobiography illustrates the act's schizophrenic nature: "One southern senator asked me, 'How'd you like that Telecom Act?' 'I've studied it a lot,' I said. 'Then you know we put everything in there. Then we put its opposite in.' And he laughed, slapped me on the shoulder, and walked away, still chuckling." Hundt, *You Say You Want a Revolution?*, 177.

14. Some of the more high-profile bankruptcies included McLeod, NorthPoint, Covad, RhythmsNet, and XO Communications. In all, the Association for Local Telecommunications (ALTS) estimates that about 50 CLECs went bankrupt between 1996 and early 2003. Some CLECs had limited success in high-density metro areas. And large existing companies such as AT&T and MCI WorldCom took advantage of provisions of the Act to get into the local telephony business, with some degree of success. For more information, see ALTS, "Progress Report on the CLEC Industry," October 2002, available at <http://www.alts.org/Filings/101702CLECProgressReport.pdf> (accessed 12 February 2003).

15. The mechanism of cross-subsidization originated in a 1930 Supreme Court decision (*Smith v. Illinois Bell*) that allowed telephone network operators to reallocate costs based on the way the network was used instead of how it was deployed. In an effort to promote universal service, long-distance rates subsidize local rates, business rates subsidize residential rates, and urban rates subsidize rural rates for basic telephone service.

16. Cream skimming refers to an industry-wide practice of targeting an incumbent's best customers with a comparable and often identical lower-cost product. Cost savings have traditionally been the result of favorable regulatory treatment that allowed new entrants to avoid certain fees or set prices at rates lower than incumbents. Cream-skimming opportunities have generally been short lived, as increasing competition leads to the reversal of the regulatory action or the market returns to more of an equilibrium by another mechanism.

17. Telecommunications companies that did find success cracking local markets were established giants such as AT&T and MCI WorldCom. They could afford more patience than start-ups and had more assets to leverage.

18. While there is a natural progression to standardization over time, a competitive marketplace is the most powerful force to encourage a product to improve and to create situations that give rise to modularity and create ability for new entrants.

19. In early 2004, power companies also began experimenting with technologies that allowed them to offer high-speed access to the Internet over the electrical infrastructure. They still face significant technological hurdles before they can be considered viable competitors. Furthermore, similar to the cable operators dis-

cussed in chapter 10, the success of power companies will depend on whether they employ a strategy that is disruptive relative to the incumbents. A sustaining strategy would offer them a slim chance of success.

20. Michael Powell, remarks at the National Summit on Broadband Development, Washington, DC, 25 October 2001. Available at the FCC web site, <http://www.fcc.gov/Speeches/Powell/2001/spmkp110.html> (accessed 18 January 2004).

PART II

Illustrations of Theory-Based Analysis

THE PREVIOUS SECTION should have helped you to build intuition about how to use the theories of innovation to predict industry change. This section furthers that intuition by showing how to use the principles in Seeing What's Next *to peer into the future of five distinct industries—education, aviation, semiconductors, health care, and telecommunications—and to analyze innovation overseas.*

Everyone who cares about any industry's future—from senior executives to mutual fund managers to policymakers to Wall Street stock analysts—must answer the same basic questions: Which innovations are threats, which innovations are opportunities, and which innovations will be duds? Which firms are best positioned to capitalize on industry growth? How will industry structure change? What choices are firms making that will help them succeed? What choices are firms making that will lead them toward failure?

To answer these questions, analysts typically rely on careful observation and data collection. They interpret a company exceeding projections as a sign of a positive trend and ratchet expectations up; they interpret a company coming in below projections as a sign of a negative trend and lower their expectations. Stock prices swing wildly in reaction to news. Is the future really this volatile?

For example, on July 28, 2003, Atlantic Coast Airways (a regional airline) announced it was ending a fourteen-year arrangement with United Airlines to go it alone as a discount operator. Over the next forty-eight hours, investors, reasoning that the loss of a large client is never a good thing, pummeled the stock. It dropped by one-third in two days. But could Atlantic Coast's decision actually be good news? How can you tell?

Theory helps us separate signal from noise. It helps us understand where to look and what to look for. It helps us understand which firms are sowing the seeds of their own destruction and which firms are planting the seeds of future growth.

The five industry-specific chapters in this part show how to use good theories of innovation to explain seminal events in an industry's history, to illuminate signals of change that could herald sweeping transformation, to evaluate looming competitive battles, and to analyze strategic choices that will determine how those changes unfold. Chapter 9, "Innovation Overseas," shows how theory can also help analyze factors that affect an entire economy's ability to grow.

We won't use all of the theories of innovation from the previous section with equal weighting in each chapter, because different theories are more or less salient to an industry's future depending on its economics and stage of development. We hope readers come away with a deeper intuition of how to apply these ideas to structure an analysis of whatever industry, innovation, or topic matters to them.

CHAPTER FIVE

DISRUPTIVE DIPLOMAS

The Future of Education

Do the theories of innovation apply to education? What is a nonconsumer in the education context? Can educational providers overshoot their students? What sorts of mechanisms can organizations use to reach nonconsumers and overshot customers? What sorts of developments could prove threats to leading universities? What would signal that the game is changing? How can universities respond? How do nonmarket players affect the industry? Can innovation help fix our ailing public school system? What do the theories of innovation have to say about the future of education?

From phones to planes to Ph.D. programs, the tools in *Seeing What's Next* provide insight into the future of many industries. This chapter turns the lenses of theory onto the education industry, with a primary focus on higher education.[1]

Many think of innovation as antithetical to education. But important innovations are reshaping the delivery of education. In 2004, for-profit competitors offering distance learning via the Internet, corporate universities, and some community colleges were growing robustly.

All of these developments bear the hallmarks of disruption. Online learning offered by for-profit providers such as the University of Phoenix and Concord Law School create opportunities for thousands of previously nonconsuming adult learners to obtain degrees in their spare time. Even while traditional business management programs churn out thousands of Master of Business Administration (M.B.A.) graduates each year, corporate universities such as GE Crotonville, Motorola University, and IBM's management training programs are teaching workers and managers the specific and remarkably sophisticated skills they need to solve the specific problems they face. And community colleges offer a low-cost way for students to become certified in many professions such as nursing and information technology, providing an opportunity to bypass a more expensive four-year institution.

This chapter begins by showing how to identify nonconsumers and overshot customers in higher education. Then we discuss in more detail how new players have emerged to reach these customer populations. Next we will show how even the most venerable institutions could face pressure as disruptive innovations harness asymmetries and move upmarket. This could be bad news for universities that don't respond properly but *great* news for consumers of education services. Disruption is the fundamental mechanism that makes products and services inexpensive, convenient, and simple to use. When these waves of innovation sweep through a market, people invariably consume more. Disruption is already well under way in higher education, broadly defined. Simplicity, convenience, and lower costs are enabling more people to consume more education. Social and economic welfare will continue to increase as more people learn at all educational levels. Disruptive business models will fling open the doors of quality education to previously nonconsuming populations.

We will end the chapter by taking a brief look at the problems plaguing our public primary schools. Many consider the public sector immune from the forces of innovation. But new-market disruption is beginning to flower, even at the lowest level. Schools could incrementally and rationally outsource more and more of their courses to private providers. One day we might wake up and see a public school system that looks a lot different than it did in the past.

Signals of Change: University of Phoenix, Concord Law, Corporate Training, and Community Colleges

In early 2004, signals of change abounded. Three important ongoing developments were the growth of for-profit providers, corporate training, and community colleges. All of these innovations latched on to the populations favorable to disruption—nonconsumers and overshot customers. So we begin by showing how to identify these groups in higher education.

Signs Point to Numerous Nonconsumers and Overshot Customers

Nonconsumers lack the time and resources to get the education that would help them solve problems that arise in their daily lives. Some people who don't consume higher education would likely never prioritize additional education—it just wouldn't help them do what they need to do. These people do not meet our definition of nonconsumers, because they don't need or want to learn new things or new skills. But of the 40 percent of Americans who do not go on to any form of higher education, or of the even larger percentage who do not go on to some form of graduate school, a significant number have problems or obstacles in their lives to which additional education could provide solutions.[2] However, they have not been able to pursue this education because they have lacked the time, financial resources, or skills (test scores or past experience).

For example, many businesspeople would love to receive customized training to help them advance in their careers. But they might not have the requisite background to get into the M.B.A. program of their choice. Or they might not have the time to attend an intensive program. These are addressable nonconsumers who could be reached with a new-market disruptive innovation.

In chapter 1, we noted that new-market disruptive innovations make it easier for nonconsumers to do something they historically prioritized. Chapter 3 of *The Innovator's Solution* explored this topic, showing how customers hire products to get important jobs done. The jobs-to-be-done theory holds that products are successful when they connect with a circumstance. We live our lives in circumstances. During the course of the day, problems arise and we look around to hire products to solve those problems. Products that successfully match the circumstances we find ourselves in end up being the real killer applications. They make it easier for consumers to do what they were already trying to accomplish.

By identifying what jobs people really care about and developing products that make it easier to achieve these jobs, companies can identify new markets they were previously unaware of that could not be uncovered via traditional market research. Understanding those jobs that are not adequately satisfied by current products provides deep insight into what are and what will be the truly innovative products that delight existing customers and attract new customers from the sidelines of nonconsumption.

How does this concept apply to higher education? Many nonconsumers have a job they need to get done beyond just "to learn." Learning is *always* important, but it is often a component of a broader job: "Let me be successful." Depending on where a person is in his or her life, that job could take the form of:

- "Get me a job."
- "Help me solve a problem."
- "Make me qualified to get the next promotion."
- "Provide certification that I have attained a certain level of education."
- "Brand me in a way that enhances my long-term career potential."

Nonconsumers welcome innovations that make it simpler, more affordable, and more convenient for them to get these sorts of jobs done.

There are nonconsuming contexts as well. Most universities continue to deliver education face-to-face, in inconvenient, centralized settings. Whenever people can't consume education when and where they want it, they are in a nonconsuming context.

That nonconsumers and nonconsuming contexts exist should not come as a surprise. But what about the other signals of disruptability—overshot customers who present opportunities for low-end disruptive innovations, modular displacements, and shifts closer to the end consumer?

There are some signs that postsecondary schools might have overshot the needs of at least *some* potential students. Let's assume that tuition maps to the performance offered by colleges and universities. Between 1980 and 2000, tuition at our country's four-year colleges and universities increased almost 8 percent annually—more than double the rate of inflation over that period. Enrollment in four-year programs has grown at a snail-like pace of .5 percent over the past decade.[3]

As in other markets, it is safe to assume that in higher education, what most students are looking to get out of school has remained relatively constant over time. Therefore, whatever innovations are driving the steady increases in tuition, it means that conventional colleges and universities are offering too much to some students. What extra functionality might be unnecessary? Some students might not care if their school has a football team. Others might not want a medical facility or a chemistry lab. Still others might not care about impeccably manicured lawns and ornate buildings.

Some of the most demanding students might want these extras. In fact, these extras might be necessary for the jobs demanding students hire universities for, such as socialization with a diverse student body or the opportunity to experiment away from parental oversight. Some want the name of the most reputable institution possible on their résumés to enhance their personal brands and improve their chances of getting the jobs they want. A few might actually want to learn as much as possible from the best people in each discipline. Topflight universities would not be overshooting these students. But increasing grumbling about tuition hikes and the growth of simpler, lower-cost options (discussed below) indicate that schools overshoot some students.

Similarly, professional schools would seem to overshoot some students. Schools' comprehensive, architecturally interdependent curricula provide services that overshoot the needs of many would-be businesspeople, lawyers, doctors, or engineers. Consider M.B.A. programs. The best general management M.B.A. programs are highly integrated. The premise is that future managers can't understand marketing unless they study product development, and they can't study product development unless they simultaneously study manufacturing. They can't learn to manage manufacturing unless they study cost accounting, and they can't study cost accounting unless they study organization design, because all these activities are inextricably related. The job of a general manager is so comprehensive and interdependent, the belief goes, that students must study *all* of its dimensions in order to understand *any* of its dimensions. Hence, the leading schools of management are extensively integrated.

This approach has three problems. First, companies that pay employees to attend intense two-year or executive education courses find that their employees don't necessarily develop the skills they need. The multidisciplinary, integrated approach at top schools gives these companies one-size-fits-all graduates who have learned things they might need to know in their third or fourth assignments after graduation. But many of these graduates need significant further on-the-job training to acquire the skills to help their employers solve their pressing needs. Even short courses that teach broad principles can provide too much breadth and too little depth. What working people often need is some way to learn exactly what they need to learn, when they need to learn it.

Second, an interdependent approach requires expensive, intensive education that takes people out of the work force. Corporations that sponsor the education of their top junior managers lose their top talent for a long period. The increasing functionality and attendant increases in the cost of attending such programs mean M.B.A. graduates demand—and indeed merit—higher salaries than they did a generation ago. Many operating companies seeking new managerial talent find that M.B.A. graduates from leading schools command starting pay levels that do not fit into their salary structures.

Finally, some companies find that when they send *individuals* out to receive a management education, the *organization* itself does not benefit commensurately. When the better-educated manager returns and plugs into a system in which others have not learned to think in similar ways, the unchanged system co-opts the returning manager.

In summary, there are indicators of both nonconsumers and overshot customers in higher education. Interestingly, many of the disruptive developments in higher education we will discuss in this chapter appeal to both customer groups. They are lower cost, more convenient, and most customized, delighting overshot customers and expanding access to nonconsumers. They also bring education into a new context. The following section discusses the signals of change that point to companies using disruptive innovations to reach these customers.

For-Profit Institutions Bring Education to New Consumers in New Contexts

For-profit providers have grown by finding new ways to reach consumers whom existing universities have historically overlooked. These consumers are delighted with a relatively simple offering that makes it easier and more affordable to get done key jobs related to learning specific skills and gaining important credentials.

This section examines two leading for-profit providers: the University of Phoenix, a fully accredited university that has both traditional campuses and an Internet offering, and Concord Law School, a fully online law program offered under the mantle of Kaplan, Inc., a subsidiary of the Washington Post Company. In addition to illustrating disruptive growth, the University of Phoenix allows us to explore the principle of relativity, and Concord Law shows how companies can find success on the fringes away from barriers to motivation and ability.

The University of Phoenix: Using the Internet to Further Its Disruptive Growth

Founded in 1976, the University of Phoenix seeks to reach nontraditional adult learners not adequately served by traditional institutions. It attracts these nonconsumers by giving them a simple product that helps them get done what they are already trying to get done—gain skills and degrees to further their professional careers.

The University of Phoenix began by renting space, rather than building campuses, and offering short programs. It received accreditation, allowing its students to receive government loans, in 1978, and has grown methodically. Most of its attendees are older students (it does not accept applicants younger than twenty-three) looking to add specific skills in a convenient fashion. The University of Phoenix meets that need by providing focused classes that meet once a week for about six weeks. It offers degrees that are highly relevant to working adults. By 2004, it offered degrees in areas such as accounting, administration, business, education, human service/criminal justice, management, marketing, nursing/health care, and technology.

Because the University of Phoenix targets nonconsumers who would not or could not attend traditional institutions, the rise of the Internet was great news. Using the Internet for education (called *e-learning*) allowed the University of Phoenix to reach even more nonconsumers more conveniently and at lower cost.[4]

The University of Phoenix first offered an online M.B.A. program through a dial-up connection in 1989—well before the development of Netscape's first browser. Students simply needed a computer with a modem, a word processing program, a spreadsheet program, and the ability to send and receive e-mail. Courses cost $250 per credit. The total cost to receive a degree was about $10,000.[5] One hundred fifty students enrolled in the program's first year.

The University of Phoenix's initial online program was also relatively simple. It was asynchronous in that it did not offer real-time instruction. It enabled the university to reach more customers who either could not afford or did not have the time to pursue a degree at a traditional university. Because the alternative for these customers was no education at all, they welcomed a relatively simple product.

The University of Phoenix's disruptive growth has been astounding. By the end of 2003, it had more than 100 campuses in the United States. It taught about 150,000 students, including more than 50,000 students online. Whereas traditional schools face stiff resistance to tuition increases, the University of Phoenix has been very good at managing a booming business. Why? Because it is improving along the trajectory that matters, the one that is not yet good enough. This has made the University of Phoenix extremely profitable. In 2003, its parent company, the Apollo Group, had a net income of close to $250 million on sales of more than $1.3 billion. The Apollo Group's stock grew about 65 percent a year from its initial public offering in 1995 to early 2004. A separate tracking stock issued for the University of Phoenix online grew more than 70 percent a year from 2001 to the beginning of 2004.[6]

The way the University of Phoenix deployed technology reinforces an important notion discussed in chapter 2. Disruption is a relative concept. The same innovation can look different to different companies. E-learning by itself is neither a sustaining nor a disruptive innovation. What matters is how companies adopt e-learning. The Internet sustained the University of Phoenix's education model, enabling it to attract more adult learners and serve them more cost-effectively. Similarly, public institutions, such as community colleges, whose charter involves teaching as many people as possible, embraced the Internet as a way to reach more students more cost-effectively. Likewise, traditional institutions embraced the Internet, but used it largely to reinforce their existing model by enhancing classroom-based courses and providing students access to additional academic resources. In all cases, institutions used the Internet to sustain their current education model.

Top schools got into trouble when they tried to use the Internet to create an offering that wasn't as good as existing solutions, but nonetheless competed against those existing solutions in established applications. The difficulties of one online postsecondary program show why theory suggests being wary of companies seeking to compete against consumption.

Columbia University's initial foray into e-learning, called Fathom, struggled. At the outset, Fathom positioned itself as an e-learning channel for lifelong learners. Fathom marketed expensive nondegree courses in topics such as Shakespeare and African American studies. But its target market was already consuming education from a multitude of simple, low-cost sources such as books, television, and the Internet. Lifelong learners hired these products to keep their minds sharp, to occupy their spare time, to gain skills for a certain hobby, or to stay current on world events. Fathom did not match the job lifelong learners hired education to achieve. Its courses were expensive and inconvenient, requiring huge time commitments. Predictably, its initial strategy floundered. Fathom tried a different tactic by offering courses for undergraduates, graduate students, teachers, and lifelong learners. Again, these groups were already consuming pretty good products. But after three years and $25 million in investment, Columbia decided in 2003 to scuttle the venture and offer online courses through a nonprofit group within the university.[7]

How could Columbia fail when at the same time distance-learning programs were growing at 33 percent annually and analysts were projecting that 5 million students would be using distance learning by 2005?[8] Columbia ran into a classic trap. Companies never succeed when they offer people a *worse* way to get a job done. Companies should always seek to disrupt their competitors. They should *never* seek to disrupt their customers. When a company is seeking to help a customer get a job done, it should ask what the customer is currently hiring to get that job done. If the company's solution isn't as good, expect it to fail. The providers that succeeded in driving growth competed against nonconsumption, offering cheaper, more convenient solutions compared to other things customers could hire to get the same job done.[9]

One hurdle all of these companies must overcome is nonmarket barriers. The next example, Concord Law, shows how companies can find success by going to places where existing barriers are irrelevant.

Concord Law School: Finding Nonconsumers and Avoiding Government Barriers

As of 2004, Concord Law was the country's leading fully online law school. Like many private players in the education industry, it had to decide early in its development how to deal with powerful nonmarket players such as accreditation agencies and industry associations. Predictably, nonmarket forces simultaneously affect both motivation and ability (see "Nonmarket Players in Education").[10] Stand-alone e-learning providers

Nonmarket Players in Education

Many of the regulations that limit innovation are based upon older educational models established for an agrarian society. Historically, the government pushed to maintain the integrity of the educational system by promoting time spent in a classroom with an instructor and discouraging "degree factories" that gave students a rubber stamp without imparting any knowledge.

The most important way the government influences the industry is through its control over a huge pot of money. In 2000, the government doled out close to $40 billion to postsecondary institutions. Provisions of the landmark 1965 Higher Education Act, whose purpose was "to strengthen the educational resources of our colleges and universities and to provide financial assistance to students in postsecondary and higher education," established the government as an important funding source for students pursuing higher education. By 2000, more than 50 percent of all postsecondary students received some form of financial assistance. Federal and state governments provided more than 70 percent of all financial aid, giving close to $70 billion in grants, loans, and work-study programs.

The 1965 act established an important role for third-party accreditation agencies, which ensure that institutions meet certain standards. Only students who attend accredited institutions have access to federal financial aid. Most accreditation programs historically had inappropriate metrics for online schools, posing a powerful barrier to an entrepreneur's ability to create an e-learning solution.

Additionally, provisions of the 1965 act deny federal financial aid to students who attend colleges that enroll more than 50 percent of their students at a distance. The intent of the legislation was to control so-called diploma mills that would provide little of value but a piece of paper claiming a person was a graduate. Similarly, the education department required students to have at least twelve hours of course work per week to qualify for federal financial aid.

Finally, regulations that affect the ability to utilize teaching materials limit the ability of entrepreneurs to reach customers. For example, the Copyright Act of 1976 provides guidelines for the use of dramatic and literary work. Under current law, only nondramatic literary and musical works can be used in online courses without permission.

find it hard to get accredited. This makes it difficult for them to tap into government loan programs, dampening their motivation. These factors complicate innovation in education.

Although people keep promising that reform is coming, government and large agencies change slowly. Often these changes are ex post facto accommodations of fait accompli changes instead of truly proactive reform. Companies have two choices in such circumstances (see figure 5-1). The first is to expend significant resources developing strong lobbying capabilities in hopes of turning regulation in their favor. The second is to see regulation as an *opportunity* to enter a market's fringe. An old adage comes in handy. Companies have been given lemons. They can protest. They can complain. They can try to return the lemons. They can hurl the lemons away in an angry fit. Or they can make lemonade.

For example, initially Concord's students were only eligible to sit for the California bar exam. American Bar Association requirements made

FIGURE 5-1

Strategies for Dealing with Nonmarket Forces

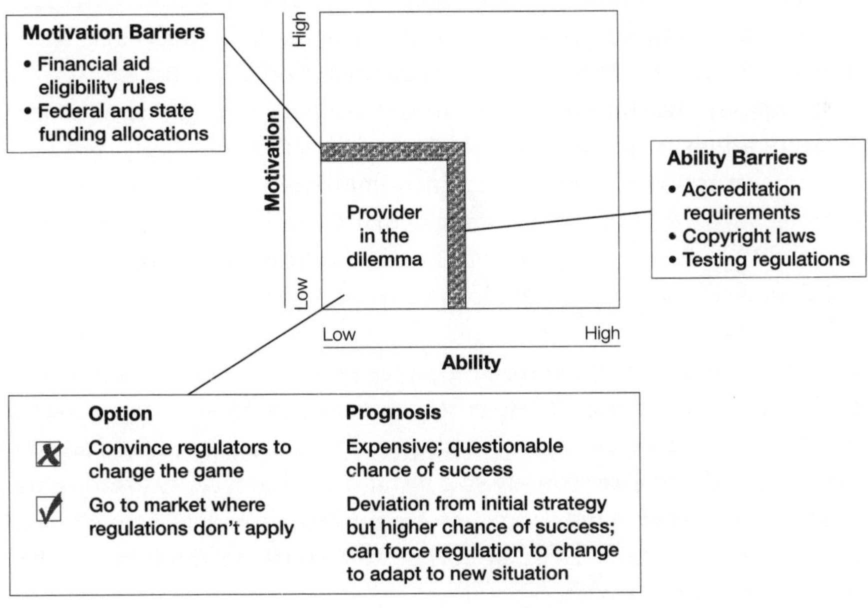

Concord's students ineligible to sit for the exam in other states until they passed the California bar. In contrast, students who graduated from traditional accredited law schools were immediately eligible to take the bar exam in any state. If you wanted to be a trial lawyer in Massachusetts, the thought of not even being able to sit for the Massachusetts exam might have been enough to steer you away from Concord.

So what could Concord have done? It could have tried to lobby. That approach would have been expensive and would have subjected Concord to the whims of industry bodies that have little incentive to change the regulations. Instead, Concord has followed a classic strategy of new-market disruption. The school recognized that it could tap into a market of nonconsumers who were seeking to achieve a different job than "help me be a lawyer." Many people want to study law but don't necessarily want to *practice* law. They want to understand how the law works to further their careers. The ability to take the bar just doesn't matter to these consumers.

The strategy appears to be working. Concord has grown from 33 students in 1998 to more than 1,000 students in 2003. More than 40 percent of Concord's students hold one or more advanced degrees. The student body includes CEOs, physicians, bankers, engineers, entrepreneurs, and stay-at-home parents. Concord serves classic nonconsumers, students who, due to geographic location or family, work, and travel obligations, would otherwise not be able to attend law school.[11]

Companies frustrated by government barriers similarly should seek to target a market segment in which the barriers don't apply and take root in disruption. As the companies improve and are able to serve wider groups of people, customer demands may force regulation to change. We believe this approach has a higher long-term chance of success than trying to force regulations to change directly.

Institutions such as Concord Law School and the University of Phoenix have developed low-cost, convenient ways to reach nonconsumers and overshot customers. Although they have innovative features, they look and feel directionally like a traditional university. Traditional universities face pressure from less traditional providers as well. The next section looks at the rise of corporate training to show how it has created disruptive growth.

Corporate Training Can Provide Customized Education Solutions

Corporate training programs have increased meteorically in recent years as they have provided convenient ways for employees to learn strategies to solve specific problems. Although precise estimates are difficult to obtain, reports suggested that the overall market size ranged from $10 billion to $30 billion in 2003. There are a reported two thousand corporate universities in the United States (compared to just four hundred in the beginning of the 1990s).[12] Given corporations' growing commitment to linking management development to corporate strategy and the ever-increasing ease and quality of educational content delivered via the Internet, as well as a captive audience of millions of managers, corporate training poses a potential disruptive threat to traditional M.B.A. programs.

In recent years, many companies have invested millions in training. Some have created their own universities and institutes that offer a wide range of certificates, degrees, and diplomas. In 2001, General Motors's (GM) corporate university provided almost two hundred thousand student days of education through approximately 1,500 courses to the company's 86,000 managerial, executive, professional, and technical employees.[13] The amount of education provided by GM alone in one year is roughly equivalent to teaching all first- and second-year M.B.A. students at the Wharton School of the University of Pennsylvania during the same period. General Electric reportedly spent more than $1 billion on its management training, much of which was delivered at its fifty-two-acre Crotonville campus nestled in New York's Hudson Valley. IBM spent more than $500 million on training modules delivered to managers in a just-in-time fashion so that they could learn exactly what they wanted, when they wanted. It recently received a new charter to become a profit center, marketing its programs to executives in other companies. Nypro Inc., a precision-injection molding company in Massachusetts with nine thousand employees, sponsors the Nypro Institute, a Massachusetts-accredited school that offers high school equivalency, English as a second language, A.A., B.S., M.S., and M.B.A. programs.[14]

How do these developments create disruptive growth? Historically, employees seeking management training had to go to two-year M.B.A. programs, or more recently to shorter but still expensive executive M.B.A. programs. Only a select few could attend. Many consumers lacked either the skills (grades, test scores) or the resources (time and money). Corporate universities open the world of management education to thousands of employees who would never consider studying for an M.B.A. degree but need access to many of the lessons found in such programs.

Furthermore, corporate training can provide more customized solutions. Traditional business programs excel at training managers in general business theory and exposing them to a diverse network of business leaders. But they are ill equipped to provide learning customized for an individual company or individual employees. Corporate training programs lack campuses or access to a high-powered alumni network outside of the company. But modular, customizable corporate training has an advantage that interdependent M.B.A. programs can't match—a product specifically designed for each employee's needs. For example, managers can take a three-day, customized course on strategic thinking; then they can go off and use what they learned to craft a better strategy for their organizations. Later, they might take a week's course on new product development and then figure out how to improve their development processes, and so on. In contrast to the leading schools' integrated structure, the on-the-job management education industry is a disintegrated one. Hundreds of specialized firms develop materials, others design courses, and others produce and teach them.

The simplest of products can delight employees who welcome the opportunity to learn skills that are directly applicable to the problems they face. Evidence suggests that managers derive more value from these programs, which relate directly to challenges they face in their jobs, than from generic two-year M.B.A. programs. This is not surprising. Corporate universities have greatly enhanced people's abilities to solve specific on-the-job problems. Prior to the rise of corporate universities, a manager facing a specific problem could not hire education to help solve her problem. She could either wade through the problem herself or hire expensive management consultants.

Corporate training has delighted millions of nonconsumers and potentially overshot customers. Again, the Internet facilitates corporations' efforts to develop robust training offerings. Companies can customize online content and access it at very low costs. In addition to internal corporate efforts, for-profit companies such as Thompson and myriad specialist consulting firms are driving corporate training's growth.

Similar to the for-profit providers discussed previously, corporate training enables nonconsumers while offering new benefits, such as convenience. Our final example shows how community colleges can be a low-cost way to deliver good-enough education, particularly in areas where schools can deliver knowledge in a rules-based way.

Community Colleges Bring Low-Cost Solutions to Overshot Customers

Community colleges are a low-cost education model. Throughout the country, these schools offer educational opportunities to students whose needs are overshot by traditional four-year institutions. Community colleges do not offer the socialization experience or range of degrees found at four-year institutions. Students hiring a community college do not need these extras to achieve the jobs they seek. Community college students attend a community college to try out college, develop proficiency in a specific subject area, or improve their academic résumés to improve their career possibilities. Community colleges provide an education at a lower cost per student than four-year institutions because their faculties are paid to teach, not to do research and publish. Hence, their overheads are lower. They offer a less expensive, more convenient alternative that squarely fits with these students' needs.

Students are increasingly turning to community colleges to get basic course credits before transferring to four-year institutions. Arizona's ten community colleges grew almost 10 percent a year from 2000 to 2003. Enrollment in the Lebanon, Pennsylvania, campus of the Harrisburg Area Community Colleges shot up 31 percent in 2003. Elgin (Illinois) Community College's student population swelled by 15 percent in 2003.[15]

A useful way to show how community colleges can drive disruptive growth is to look at developments in nursing education. Over the past forty years, the Associate Degree for Nursing (A.D.N.) program has emerged as a low-cost, accessible, and recognized education alternative for becoming a registered nurse (R.N.).

In the 1950s, there was a critical nursing shortage. In her dissertation for Columbia University Teachers College, Mildred Montag found that existing four-year Bachelor of Science in Nursing (B.S.N.) programs provided too much for many students interested in a nursing career.[16] R.N.'s had to possess a defined set of knowledge to pass state licensing exams. A B.S.N. program provided that basic knowledge along with extras such as nursing theory and liberal arts education. This was overshooting.

Based on Montag's work, schools developed A.D.N. programs focused on just the basics. A.D.N.'s were shorter (two years rather than four), simpler, and cheaper. Community colleges could provide A.D.N. programs in a lower-cost, streamlined way. A.D.N. programs quickly became the primary vehicle for educating R.N.'s. The number of A.D.N. programs skyrocketed from 7 in 1958 to more than 850 by the mid-1990s. Today, more than 60 percent of all R.N.'s are graduates of A.D.N. programs.[17]

Although there is a significant difference in cost and time commitment, there is no conclusive evidence to suggest that a four-year B.S.N. degree better prepares one to be a nurse than the less expensive two-year A.D.N. According to a professor at a prominent northeastern four-year program, "B.S.N. nurses have more academic knowledge than others and are more aware of the theory, the history of nursing. But clinically [the two types] are probably equal."[18] Although an A.D.N. is not good enough for nurses who wish to work in specialized fields or pursue advanced degrees, it is more than good enough for people seeking to become R.N.'s.

Signals of Change Abound

All of these innovations have the potential to change higher education. For-profit providers have grown by making it easier for nonconsumers to get the education they desire. Corporate training lets managers learn what they want, when they want it without going to an expensive M.B.A. program. Community colleges provide a low-cost vehicle for teaching the basics. All of them promise to provide more education to more people at lower costs. That is a good thing. But what about our leading universities? Are these developments potential threats? The next section looks at potential competitive battles and suggests the answer is yes.

In Upcoming Competitive Battles, Universities Should Beware

The previous section pointed to several innovative ways to deliver education that are driving industry growth. All of these innovations are improving and expanding, reaching more and more customers. Could these potentially disruptive opportunities spell the end of our traditional education system? Are asymmetries credible in education? Through the lenses of our theories, the controversial answer is yes. Asymmetric motivation continues to fuel disruptive growth. Companies are developing asymmetric skills vis-à-vis traditional universities. Some universities could feel pressure soon. Although our most distinguished schools might be safe for decades, they run the risk of missing the opportunity to drive the education agenda as new providers create unique skills around customization and convenience.

The first step in evaluating potential competitive battles is to take the tale of the tape, looking at incumbents' resources, processes, and values to assess their motivation and skills. Let's consider our very best four-year universities, such as Harvard, Stanford, and Duke universities.

Top schools have powerful resources, such as high-profile faculty and alumni, large endowments, sprawling campuses, and well-established brands. To stay at the top of their game for decades, leading universities have developed effective processes to find and nurture faculty, attract and retain students, and provide best-in-class instruction. Specific processes include things such as professor recruitment, research, faculty review, admissions selection, curriculum development, physical plant management, and so on. All of these processes provide powerful skills schools can deploy to produce world-class research and attract and educate the very best students.

Although most schools are not-for-profit organizations, financial considerations still influence their values. Schools have to pay faculty salaries, maintain their facilities, and support research activities. They therefore must charge relatively high tuition rates (or receive government funding) to cover these costs. Also, top schools want to be the best. They believe they have a mission to provide the very best education to the very best students while letting the very best researchers do the very best research. These values drive leading universities' resource-allocation decisions. Universities' not-for-profit status does lead to one important difference—top universities just don't have as much motivation to grow as for-profit companies do.

Immediately, we see two asymmetries disruptive entrants took and continue to take advantage of: differences in target customers and business models. First, the innovations discussed in this chapter target a very different student group than leading universities target. Disruptive companies find success in serving students who are unattractive to the mainstream universities (measured by life situation, test scores, grades, and so on)—a perfect asymmetry. Second, disruptive educators feature unique business models. They tend not to have large campus infrastructures or high-priced instructors. This difference enables them to be viable at price points that leading private institutions would find untenable.[19] Also, they tend to provide short, special-purpose courses that need not be knitted together in degree programs, which does not fit existing universities' education models.

These shields explain why leading institutions just haven't been motivated to go after the opportunities that delight for-profit universities, corporate training providers, and community colleges.[20] These forces suggest that while traditional four-year schools might offer open-enrollment extension courses, they will do so as a way to leverage fixed costs, not as a concerted strategy to grow. Thus, the potential disruptors have asymmetric motivation on their side, shielding them while they grow.

Are the entrants creating asymmetric skills? Again, signs point to yes. Entrants are developing the unique ability to deliver low-cost, customized, convenient education. Potentially disruptive innovators are mastering new ways of teaching. Instead of delivering multiyear, multidisciplinary programs, they are developing a set of highly specialized, shorter offerings. They are developing the unique ability to use technology to facilitate the learning process. They are learning how to virtually manage thousands of student interactions.

Most leading universities consider themselves at the top of their games. Applications for scarce spots keep rising. Students and their families continue to be willing to fork over tens of thousands of dollars for education. Now would not seem the time to worry about disruptive threats. Surely, leading schools think, there will always be a market for the integrated, in-depth learning they provide. But companies are growing behind the shield of asymmetric motivation. They are forging swords of asymmetric skills. What would disruption look like?

What Could Happen: A Slow Disruption

Because the education industry is so vast, let us focus on management education's future as an example of what the theories imply for the broader set of institutions.

First, as M.B.A. programs become increasingly expensive, more and more companies will begin recruiting their management talent directly from bachelor's degree programs and insert those managers into their corporate universities. Those corporate universities do not now, and will not for some time, compete against leading schools in the "brand-myself" job segment. But as they seek to become better at doing the "get-the-required-credential" and the "solve-my-immediate-problem" jobs, they will be motivated to get the very best materials from the very best instructors. Often this will be done via video, DVD, or the Internet.

Branding would therefore migrate from the assembler—the university—to the course instructor. It might be very important that you took a course on process redesign from the guru, Michael Hammer, but be unimportant whether you took the course at a university, your corporation, or an Internet-based offering.

It might take a long time before top schools feel any pain. The first to feel pressure would be second- and third-tier schools. Indeed, more than two hundred institutions closed their doors during the 1990s.[21] By the time it becomes clear the game is changing, it will be too late to respond. In essence, disruption would involve entrants pulling increasing groups of students into their contexts, where the leading institutions' unique strengths would be unique weaknesses. Leading institutions would be too slow, too expensive, and not customized enough to compete with the entrants. These top institutions would find themselves fighting over an increasingly small pool of the very best students who would still need intense, interactive training in the most complicated subspecialties. Ultimately, the number of traditional universities would plummet.

How Leading Schools Could Respond: Skating to the Back End or Launching a Counterattack

What could leading universities do to fend off this disruptive incursion? Flight wouldn't seem to be an option. Leading schools are already highly selective, so there wouldn't seem to be large enough numbers of undershot customers who reside further up-market. Leading schools essentially have three options: Do nothing and hope for the best; skate to the back end by attempting to profit from learning modules; or respond by attempting to grow their own disruptive offerings.

The first option involves ignoring the disruptive threat. Schools would hope that the unique jobs they provide (socialization, the stamping of the schools' brand on the graduates, and so on) mean there will always be pockets of students who demand their services. That certainly is possible. However, the viability of this strategy is contingent on the disruptive institutions *not* being able or motivated to improve.

The second approach would involve "skating" to the new spot in the value chain where things aren't good enough. Think again about M.B.A. programs. Historically, value went to integrated players. As companies change the basis of competition to convenience and customization, what is not good enough is the ability to give every student exactly what he or she needs to learn individually, exactly when he or she needs to learn it. This entails creating learning modules, with their associated instructional guides, that make it simple for corporate trainers to assemble and deliver great materials, customized exactly to each learner's needs. Today's leading schools would seem well positioned to create modules that improve upon this new dimension of what is not good enough, and, if they do, they will capture value. Leading professors from top schools already participate in corporate training. Corporate trainers use case studies and research notes from leading universities. Similar to the way Intel decided to provide value-added chip sets and motherboards rather than bare microprocessors, leading business schools could provide learning modules to become the "Intel Inside" of corporate training programs. The modules leading schools would provide would enable firms that design, assemble, and produce courses to mix and match materials to address even more responsively and effectively the education and training needs of their customers. Their goal should be to make it instantly simple for trainers to teach great material in an engaging way. If the leading schools of management worked in this way to facilitate their own disruption, they likely would find that they could fuel the growth of their industry, have vastly greater influence over how managers think, and continue to enjoy much of its profit as well.

The third strategy would be for leading schools to not just fuel the disruption, but to be the disruptors themselves and grow into new markets by creating simple offerings that let them reach a wider population. Assessing the success of this strategy would require carefully watching how schools handle the values problem that might direct their resources elsewhere. A desire to give the best to the best would run counter to bringing the good enough to the masses.

As of 2004, leading schools were experimenting with novel ways to find nonconsumers. One particularly interesting approach involves

INDUSTRY SUMMARY

Higher Education

Signals of Change: For-profit institutions, corporate training programs, and community colleges all are creating disruptive growth with hybrid low-cost/new-market models.

Competitive Battles: Growth takes advantage of asymmetric motivation by using unique methods to target different students; new players could develop unique capability to deliver customized and convenient programs.

Strategic Choices: Leading schools must decide whether or not to create their own disruptive offerings or provide modules to disruptive entrants.

---

bringing low-cost education overseas. Universitas 21, a consortium of international universities, has developed a partnership with Thomson Learning to develop Universitas 21 Global. It launched an online M.B.A. program in 2003 and plans to introduce a Master of Information Systems degree. It has set its sights on the Asian market. There is high demand for M.B.A. programs in China and other parts of Asia. The high costs and general difficulty associated with moving to another country for two years limits the number of Asian business school students. These are classic nonconsumers. Universitas 21 Global's online M.B.A. program costs about $10,000 and is available to anyone with access to the Internet. It has a much greater chance of success because it provides the means of receiving an advanced degree to thousands of nonconsumers.

Public Schools: Let Disruption Teach

Finally, let us briefly examine the problems that plague our public school system through the lenses of the theories of innovation. Performance of public kindergarten through twelfth grade (K–12) education is uneven but generally poor. Schools face never-ending streams of budget cuts. Dissatisfaction is growing. The United States spends more than any other country on primary education, but our students rank below almost a dozen other countries on international standardized tests.[22] An ideal education system would meet all students where they are and help them reach as far as they can go. There is an unambiguous consensus that our public schools don't do this well. Are there opportunities for disruption to reshape the way we deliver primary education? This section discusses two innovations that many observers believe could reshape the delivery of primary education: charter schools and computer-based learning through the Internet. Our lenses lead us to discount the ultimate impact of the charter school movement. Charter schools have played an important role in helping schools move up-market, but they seem to be set up to sustain rather than disrupt. On the other hand, using the Internet to deliver courses seems to contain great disruptive potential. It could allow a radical transformation to happen in an incremental, rational way.

Charter Schools: Set Up to Sustain

State charter laws emerged in the early 1990s.[23] They permitted individuals or groups to set up independent public schools outside the existing system. These new schools had to sign a contract, or "charter," with the state or local government establishing performance goals. Designers hoped that charter schools would be progressive laboratories of innovation because they are largely free from requirements that shackle public schools.

Education reformers aggressively opened charter schools throughout the 1990s. By fall 2002, 2,700 charter schools were operating in thirty-six states and the District of Columbia, serving more than 575,000 students. Although charter schools only account for approximately 1.5 percent of total public school enrollment, the 2002 figures are a significant increase from 1999, when there were only 1,010 charter schools educating 266,000 students. Charter schools are more likely than regular schools to be located in urban areas, to enroll fewer than 200 students, to enroll a higher proportion of black and Hispanic students, and to have a higher percentage of students eligible for free or reduced-price lunch programs. Additionally, charter schools tend to employ teachers with fewer years of teaching experience.

Recall that the reason companies should follow a disruptive strategy is to establish a market presence and growth trajectory that incumbents are unable to defend against. Instead, a disruptive strategy should entice powerful, existing competitors to flee the entrant (or ignore the entrant), rather than to fight the entrant. This is not the strategic intent of charter schools. Their purpose, as sustaining innovators, actually *is* to create a fight—to stimulate competition among public schools that will force them to innovate and improve. Many states in fact give charter schools more money to target high-need students to help pull them up the performance trajectory more effectively, it is hoped, than public schools have been able to do.[24]

In the context of primary education, this development is certainly a good thing. Public schools have struggled to improve to meet students' needs. A strong argument can be made that the public primary and secondary education systems undershoot many elements of the population. Public primary schools are in an unenviable position. They hope to provide the best education possible at the lowest possible cost, but each child has different needs. School systems are trying to simultaneously educate students that range from the potential dropout to the learning disabled to the honor student. Schools have little choice but to offer a one-size-fits-none product. The burden is left to teachers to customize the learning within the confines of publicly defined and mandated curricula. Although in most situations we would expect incumbents to successfully innovate to meet these undershot customers, system constraints (the government, local school boards, socioeconomic factors, and so on) make it very difficult for most schools to improve.

Because charter schools follow similar teaching models to existing schools, they have limited potential to disrupt—that is, to make it possible for each student to get exactly what he or she needs, when he or she needs it, at lower costs—so students consume more education.

Hope on the Horizon: Virtual Schools and Asymmetries of Motivation

Recall our discussion about how different types of higher education providers implemented e-learning in distinctly different fashions—but each in a way that sustained their current business models. Now think about computer-based learning in our public school system. Will it have a sustaining or a disruptive impact? It depends on the strategy used to implement it. Using computers in today's school classrooms to foster anything except incremental improvement with the current model has a very low probability of success. Teachers and administrators are unlikely to cede power and control. Parents and administrators will debate as to how good a job computers can do to adjust to the needs of the student. To date, schools have largely deployed computers in a cost-additive way that either complements or enhances the existing system's performance.[25]

But a new-market disruption has already begun in a stunningly classic fashion. Apex Learning (founded by Microsoft cofounder Paul Allen) has developed a product that allows school systems to offer more advanced placement (AP) courses to more students by placing the courses online. AP courses are college-level courses offered to high school students. If the students score high enough on a standardized exam at the end of the course, they receive college credit. There is vast nonconsumption of AP courses in most high schools. Many offer none. Those that do offer only a fraction of the thirty-four courses for which AP exams are available. In early 2003, Apex reported serving close to fifty thousand students in forty-eight states.[26] School systems are able to aggregate the demand for AP courses over an entire school district in subjects in which there is insufficient demand in individual schools to merit appointing a dedicated teacher.

Similarly, Concord, Massachusetts-based Virtual High School (VHS) started teaching virtual classes in 1996. By 2003, it had 120 teachers teaching 120 courses to 3,200 students in 175 schools. VHS has an innovative membership model. Schools pay a yearly membership to join a consortium. Most members of the consortium provide a course to the other consortium members. Students can take courses on topics such as the literature of Charles Dickens, AP biology, or the Vietnam War. VHS found success in rural and so-called urban fringe schools. About 80 percent of their member schools have fewer than 1,500 students and 27 percent have fewer than 600 students. An overwhelming majority of students are juniors and seniors. Given their size, these schools struggle just to provide the basic curriculum. None can afford to offer these kinds of specialized courses.

So is this strategy disruptive? There is a clear asymmetry at work. Companies such as Apex Learning and Virtual High School don't provide courses that the public schools *want* to teach. They provide courses that, given budgetary pressure, the public schools would be relieved *not* to have to offer. Schools will continue to face budgetary pressures. Year after year, the quick fix would be to ax a noncore course. Rather than deny students the learning option altogether, the school could outsource it to an e-learning provider such as Apex or VHS. The e-learning provider, in contrast, would be motivated to figure out how to teach more courses, more effectively. When the school faces more budget pressure, and another course that lacks a critical mass of students needs to be axed, the e-learning provider could say, "Hey—that previous course you outsourced worked so well. Let us do this one for you, too." The provider would be motivated to add the very course the school would be motivated to drop. And the e-learning providers would keep getting better. When budgets were cut again, the e-learning providers could offer more courses that inch closer to the core curriculum. Through a rational and incremental process, schools would outsource more and more to virtual providers. The virtual providers would grow based on the asymmetric motivation and develop unique capabilities to deliver convenient education to large groups of students. One day, the school would realize that it was teaching very few courses through traditional live instruction.

What would a disrupted world look like? Students would still go to school. What would be in the school? Banks of computers with virtual classes. Students would spend the bulk of their time doing virtual work. The school would provide music, drama, and athletics. Teachers would act as counselors, helping each student develop and adhere to a customized learning plan.[27] This arrangement would better serve the hardest-to-educate students, but it also would better serve the most motivated students. Teachers could spend more of their time working hands-on with individual students. It would cost taxpayers *much* less money. Everyone would win.

The reason a new-market disruption pathway to this transformation would work, whereas a "cram-the-computer-into-the-classroom" approach would not, is that the transformation will not require any dramatic, risky policy or strategy decisions by administrators or school committees. It only requires incremental, sensible decisions. We emphasize again that this disruption, like all disruptions, can have no transformative impact if it is initially deployed within the mainstream. The only way it can succeed is to compete against nonconsumption and then progress, one rational step after another, toward the core. If Internet educators follow a strategy like this, the result will be more access, better quality instruction, happier teachers, and lower costs.

Disruptive forces course through the education industry. New entrants such as the University of Phoenix have grown by taking advantage of asymmetric motivation. They have reached traditionally overlooked customers with relatively simple, convenient offerings. They will continue to improve and sharpen their ability to offer customization and convenience. Corporate trainers will continue to make education more accessible and relevant to employees. Community college providers will fine-tune their ability to give overshot students just the basics, and then more. In short, clear signals of change are beginning to appear that point to a different world in the future.

The confluence of new business models and technology all but guarantees continued change. It might take time for organizations to move up their improvement trajectories, but—unless something unanticipated happens—they will move up. Our very top schools might be safe because extremely demanding customers might continue to demand things that new companies just can't provide. But disruptive forces could threaten many schools unless they take action and do it soon. Remember what Western Union said about the telephone!

Similarly, organizations could use disruptive strategies to bring about a radical transformation in our public school system. Many reform efforts struggle to bring about real change because people naturally hesitate to make sweeping changes. What makes disruption so powerful is that it occurs through a series of incremental decisions, each of which appears completely rational. In other words, if providers and reformers follow disruptive strategies, radical change can happen without anyone ever making a radical decision.

The education chapter has five general lessons:

- *Nonconsumption exists, even in contexts in which everyone seems to be consuming.*
- *Making it easier for nonconsumers to do things that matter to them is the ticket to growth (for example, University of Phoenix or corporate training).*
- *Companies can adopt the same innovation in very different ways (for example, e-learning).*

- *Companies can start at the fringes and then move to the core to avoid being snared in government regulation (for example, Concord Law).*
- *All organizations—even nonprofit ones—have skills and motivation that drive what they can and cannot do.*

Notes

1. Admittedly, the education industry's complexity makes precise prediction difficult. There are too many schools to count, each with an idiosyncratic tale of the tape. Signals can be muted and difficult to observe. Nonmarket players wield tremendous influence. We hope industry participants can use our approach to undertake their own analysis of such topics. Our thanks to Sally Aaron and William Clark for their contributions to this chapter.

2. U.S. Department of Education, National Center for Education Statistics, *The Condition of Education 2003* (Washington, DC: U.S. Government Printing Office, 2003). Specific statistics available from the U.S. Department of Education Web site; see for example "Indicator 25: Educational Attainment," <http://nces.ed.gov/programs/coe/2002/pdf/25_2002.pdf> (accessed 16 February 2004).

3. U.S. Department of Education, National Center for Education Statistics, *The Condition of Education 2003.*

4. People use the term e-learning in myriad ways. In this chapter, we use e-learning to mean the technology and related academic services involved in using computers and the Internet to facilitate education. E-learning is the latest incarnation of an innovation that first appeared more than one hundred years ago—distance learning. Distance learning's origins trace back to the "mother of American correspondence study," Anna Eliot Ticknor. In 1893, Ticknor set up the Boston-based Society to Encourage Study at Home, which featured monthly correspondence with guided readings and tests. Over time, people adapted innovations such as the radio and the television for educational purposes. All of these innovations created growth but had obvious limitations. They were inefficient, slow, and not interactive. They were not ideal ways to provide collaborative, customized education. Enter e-learning. E-learning grew during the 1990s along with the Internet. Venture capitalists invested about $6 billion in e-learning providers between 1990 and 2000. Cisco CEO John Chambers claimed, "Education over the Internet is going to be so big it is going to make e-mail usage look like a rounding error." See Thomas L. Friedman, "Foreign Affairs: Next, It's E-ducation," *New York Times*, 17 November 1999.

5. Robert Bellinger, "MBA Education Goes On-line," *Electronic Engineering Times,* 24 September 1990.

6. Stock prices and financial information downloaded from Yahoo! Finance, <http://finance.yahoo.com> (accessed 24 January 2004).

7. Karen W. Arenson, "Columbia's Internet Concern Will Soon Go Out of Business," *New York Times*, 7 January 2003. Columbia was not alone. A study of online learning found that many schools' efforts to develop online learning programs failed simply because they tried "to provide traditional courses in a nontraditional manner." Many colleges attempted to develop programs using the Internet to create online courses involving two-way communication between topflight professors and their high-paying students in a remote location. They were marketed as a comparable alternative to a "bricks-and-mortar" education. As many institutions and students came to realize, however, the quality of the program did not compare to the classroom-based degree, nor did it justify the hefty tuition costs schools charged to recover their investments and to pay their professors' high salaries. Attempting to market an inferior product to their most demanding customers was, predictably, a recipe for failure because the universities were clearly cramming distance learning to fit their old business models. Moreover, universities and professors who went on to develop different types of online courses, thinking the failure was in the course design, failed again when they marketed their new courses to the same demanding customers. The problem was that they were still competing against consumption. See "Virtual College," *Salt Lake Tribune*, 5 May 2002.

8. The U.S. Department of Education estimated that the number of online courses grew from 1.34 million in 1997–1998 to 2.87 million in 2000–2001. See also Kevin Lyons, "Popularity of Online Graduate Studies Grow by Degrees," *Knight Ridder Tribune Business News*, 31 July 2003; J. M. Pethokoukis, "E-Learn and Earn," *U.S. News and World Report*, 24 June 2002.

9. Again, note that some of these people may have been "consuming." However, they had to go to centralized settings to take integrated, in-depth courses. They were nonconsumers in decentralized settings where they could take courses at a different pace.

10. Higher Education Act of 1965, Public Law 89-329, 79 STAT 1219. From the Higher Education Resource Hub, <http://www.higher-ed.org/resources/HEA.htm> (accessed 5 December 2002). Estimates for total financial aid vary. Federal statistics come from National Center for Education Statistics, "Table 362: Federal Education Support and Estimated Federal Tax Expenditures for Education, by Category: Fiscal Years 1965 to 2001," in *Digest of Education Statistics, 2001* (Washington, DC: U.S. Department of Health, Education, and Welfare, Education Division, 2002), <http://nces.ed.gov//pubs2002/digest2001/tables/dt362.asp> (accessed 5 December 2002).

11. For more information on Concord Law, see Martha Neil, "Virtual Lawyer," *ABA Journal*, December 2002; Dan Carnevale and Florence Olsen, "How to Succeed in Distance Education," *The Chronicle of Higher Education*, 13 June 2003; "World's Only Online Law School Proving Its Mettle, " *PR Newswire*, 14 August 2002.

12. Michael Brennan, "U.S. Corporate Business Skills Training Forecast Update, 2003–2007," Report 29825, International Data Corporation, July 2003;

William C. Symonds, “A New Push to Privatize,” *BusinessWeek*, 14 January 2002; Linda Anderson, “Survey-Business Education—Rivalry? No . . . It Is Really a Case of Synergy,” *Financial Times*, 25 March 2002; Linda Anderson, “Training Programmes Are No Longer the Sole Territory of Universities,” *The Banker*, 1 May 2002.

13. Linda Anderson, “Powerful Drive to the Top League,” *Financial Times*, 27 May 2002; Bureau of Labor Statistics, *2000 National Occupational, Employment and Wage Estimates: Management Occupations*, <http://www.bls.gov/oes/2000/oes_pub2000.htm> (accessed 21 January 2004).

14. Nypro, Inc., “FAQs,” on Nypro, Inc. Web page, <http://www.nypro.com/company.html> (accessed 12 August 2002).

15. Jamilah Evelyn, “Many Community Colleges Report a Boom in Their Enrollments,” *Chronicle of Higher Education*, 19 October 2001.

16. Patricia T. Hease, *The Origins and Rise of Associate Degree Nursing Education* (Durham, NC: Duke University Press, 1990).

17. U.S. Department of Health and Human Services, “Projected Supply, Demand, and Shortages of Registered Nurses: 2000–2020,” July 2002, <http://bhpr.hrsa.gov/healthworkforce/rnproject/report.htm#chart2> (accessed 18 August 2002), and National Organization for Associate Degree Nursing, “Associate Degree Nursing (ADN) Facts,” <http://www.noadn.org/adn_facts.htm> (accessed 12 August 2002).

18. Melinda M. Karp, “Nurse Education and Practice: What We Know and What We Need to Know,” (paper presented at American Association of Community Colleges, New York, April 2002), <http://www.tc.columbia.edu/ccrc/Presentations/AACCpresentation.MMK.pdf> (accessed 12 August 2002).

19. In 2003, the University of Phoenix charged about $325 per credit hour. This made it significantly cheaper (from a student perspective) than four-year private universities, but more expensive than government-funded four-year public universities. Because the University of Phoenix largely reaches nonconsumers, its relative cost position is not crucial to its success, but it will be important for its up-market march.

20. Of course, some lower-tier schools might be extremely motivated to go after these sorts of opportunities. This could potentially put a cap on disruptive growth but would provide a further threat to the very best universities. The principle of relativity helps us look at this type of question.

21. National Center for Education Statistics, “Degree-Granting Institutions That Have Closed Their Doors, by Control and Type of Institution: 1960–61 to 2001–02,” <http://nces.ed.gov/programs/digest/d02/tables/dt245.asp> (accessed 16 February 2004).

22. OECD Programme for International Student Assessment (PISA) Database 2001, table 3.6, available at <http://www.pisa.oecd.org/knowledge/annexb/t3_6.htm> (accessed 21 January 2003). This considers a country’s mean performance on PISA’s math, science, and reading evaluations.

23. Sources for this section include “The Center for Education Reform,” <http://edreform.com/pubs/chglance.htm> (accessed 20 October 2003); National

Center for Education Statistics, *The Condition of Education 2002* (Washington, DC: U.S. Department of Education, 2003), <http://www.uscharterschools.org/pub/uscs_docs/gi/overview.htm> (accessed 10 February 2004).

24. At some level, it is not surprising that charter schools have followed a sustaining strategy. Because education is freely available, it is conceptually difficult to imagine elementary and secondary schools overshooting our children's needs. Of course, education is not free to society. But individual consumers of public education don't face an incentive to opt for a not-as-good-but-cheaper solution. So low-end disruption is neither practical nor possible. Furthermore, charter laws obviously influence the strategy a charter school selects.

25. Research by Harvard Business School professor Frances Frei suggests this is usually the case. When organizations add new technologies—even technologies that purport to lower costs—they tend to do it in a cost-additive way. This is because the organization must keep utilizing its higher-cost, existing technologies for a long time. So even though the marginal cost of using the new technology is lower, it ends up being cost additive. For example, Frei's research showed that automated teller machines actually were cost additive because banks had to keep their branches open to meet customer demands. See Dennis Campbell and Frances Frei, "The Cost Structure and Customer Profitability Implications of Electronic Distribution Channels: Evidence from Online Banking," unpublished working paper, Harvard Business School, Boston, 2002.

26. Apex Learning, <http://www.apexlearning.com/about/default.asp> (accessed 13 February 2003).

27. Indeed, this model is emerging to serve students whose athletic demands make it difficult for them to fit within the existing school system. Instead of having to fit athletic events around an academic schedule, they fit an academic schedule from a virtual high school around their athletic schedule.

CHAPTER SIX

DISRUPTION SPREADS ITS WINGS

The Future of Aviation

Which aircraft manufacturers have the edge? Does the future belong to the Airbuses and Boeings of the world or the Embraers and Bombardiers of the world? Why do the major airlines find it so difficult to sustain acceptable profits? Are they just not innovative, or is something else going on? Why has Southwest Airlines been so successful? How high can discount airlines fly? What future do regional airlines have? What sorts of innovations could dramatically reshape the aviation industry?

Airplane flight is a modern-day miracle. Every day, tens of thousands of flights move people and goods around the world at speeds unimaginable a century ago. This chapter continues our industry-specific analysis by peering into the future of the aviation industry's two principal market segments: aircraft manufacturers and commercial airlines. In brief, our assessment is that:

1. Regional jet manufacturers pose a legitimate disruptive threat to the industry's leading manufacturers, Boeing and Airbus.
2. Discount airlines are unlikely to fundamentally reshape the industry because it is hard for them to hide behind a shield of asymmetric motivation; ironically, "flight" is not an option for major airlines.

3. If regional jet airlines—or next-generation air-taxi providers—break free from the grasp of major airlines, they could radically reshape the industry.

In this chapter, we first show how our theories explain important historical developments. We point to potential signals of change, evaluate looming competitive battles, and identify critical strategic choices that will determine whether disruption does in fact spread its wings.

Developments in Airplane Manufacturing: From Boeing to Bombardier

We begin by looking at the manufacturers who actually design and build the planes. This section explains how a duopoly emerged for the most-demanding customers, how two firms created a new market away from those powerful incumbents, and why those disruptive entrants could have the edge when a battle emerges.

Major manufacturers sell their products to airlines. Historically, the most demanding customers were major airlines that needed planes capable of safely taking hundreds of passengers thousands of miles. Major airlines also needed somewhat smaller planes that traveled hundreds of miles. At the lower end of the market were smaller airlines that only needed small planes to take handfuls of passengers short distances.

Meeting customer needs is a tricky task. Airplane manufacturing is technologically complicated, and financially risky. Consider the investment required to build a next-generation aircraft (in our language, a radical up-market sustaining innovation).[1] In the early 1990s, Airbus set its sights on the highest tier of the market. Boeing historically owned the market for very large, long-haul aircraft with its 747 and 777 planes. To fight Boeing, Airbus developed the A380. Analysts likened the A380, which Airbus will begin producing in 2004, to an airborne cruise ship. It seats more than five hundred people and retails for a cool $200 million. Airbus can modify the plane to include amenities such as sitting lounges or a restaurant. Developing the A380 took about a decade and cost about $10 billion. It required designing and closely coordinating a far-flung supply chain comprising literally hundreds of suppliers. Given the diverse production locations, Airbus had to invest in things such as new roadways and other infrastructure improvements to enable parts of the fuselage to reach final assembly. This is not a game for the faint-hearted.

Airbus's effort to fight for the market's highest tier is consistent with its history. When it was deciding how to enter the industry in 1970, it encountered the same basic set of choices facing all would-be entrants. It could try to bring a better product to an established market; it could seek to reach nonconsumers; or it could try to bring new benefits to existing customers. Airbus chose the first option.

At one level, this decision was predictable. Airbus naturally eyed the largest market, where success would bring the most prestige. Boeing, McDonnell Douglas, and Lockheed enjoyed a comfortable triopoly at the industry's highest end. The market looked lucrative. The predictable decision carried predictable risks. Airbus almost immediately found itself in the crosshairs of large companies that had a lot to lose. The only way Airbus could succeed was to have deep-pocketed investors who could tolerate large losses. Fortunately, with close ties with many European governments, Airbus did. Slowly but surely, Airbus made inroads into the market. As the disruptive innovation theory would predict, that success came at a stunning cost. Experts peg government subsidies to Airbus at about $30 billion.[2] The market barely had been large enough to sustain three competitors, let alone four. Airbus was able to outlast Lockheed and McDonnell Douglas. Lockheed exited the commercial aircraft market in the 1980s. Then Boeing acquired McDonnell Douglas for about $14 billion in 1997. It emerged as a behemoth that could sumo wrestle Airbus for world aircraft domination.

Looking at the tale of the tape, both companies' skills and motivation are in alignment. Both companies make money in similar ways—selling a limited number of very high-priced aircraft to large carriers. Both create incremental variants of platform products (for example, the Boeing 737-200, -300, -400, and so on) with each variant offering increased range and passenger capacity. Historically, they both prioritized investments that would help them earn the bragging rights associated with being the primary aircraft supplier to the world's premier full-service carriers. They both have honed the ability to manage complex product development and a complicated supply chain. But looking for signals of change quickly focuses attention on the disruptive growth Embraer and Bombardier have created in a completely different market.

Embraer and Bombardier Created Growth in New Markets

While Boeing and Airbus remain locked in an up-market race from which neither dares withdraw, two smaller companies have quietly created new markets.[3]

Instead of fighting Boeing and Airbus on their turf, Canadian manufacturer Bombardier created the regional jet market in 1992 when it made the first delivery of its Canadair Regional Jet (CRJ, later renamed the CRJ-200). The first regional jets offered by Bombardier and Brazil-based Embraer were limited.[4] They only had about thirty seats. They couldn't fly more than five hundred nautical miles, about the distance from Washington, D.C., to Atlanta, Georgia.

Given these limitations, regional jet manufacturers would have failed if they had initially tried to compete against consumption by going after Boeing's most lucrative customers. Customers would have considered the early versions of their products inferior. The manufacturers would have raised the ire of extremely deep-pocketed companies who had a lot to lose.

Instead, Bombardier and Embraer found a happy niche serving smaller regional airlines. (We will discuss the growth of regional airlines in depth later in this chapter.) A regional airline's only prior option was to purchase a turboprop plane that was noisy and slow and could only fly a very limited distance.[5] A series of weather-related incidents also created public perception that turboprop planes were not well suited for cold, icy climates. These jets enhanced regional airlines' ability to get people to major airports from out-of-the way locations such as Lebanon, New Hampshire; Ithaca, New York; or St. George, Utah.

Interestingly, the overall increase in air traffic allowed Boeing and Airbus to sell even more of their large jets. More people could get from outlying airports to central hubs. Demand for longer routes and the planes that flew those longer routes grew.

This growth was a signal of change. Regional airlines had been nonconsumers of jet aircraft. Embraer and Bombardier found a way to reach these nonconsumers with a relatively limited product. They created explosive growth in a different market with a product that the mainstream market found too small to use. They created this growth hiding behind the shield of asymmetric motivation. Large incumbents weren't interested in these emerging markets. The prospects of selling small planes at low prices couldn't compare to the option of selling large planes at high prices.

Regional Manufacturers Could Have the Edge in Competitive Battles

Bombardier and Embraer knew the rewards would be great if they could increase the size and range of their jets. So they improved. In 1995, Embraer (the ERJ-145) and Bombardier (the CRJ-200) sold planes with capacities of about fifty seats for $15 million to $20 million. By 2001, Bombardier introduced its CRJ-700 with seating capacity of about seventy. Embraer planned to introduce its own seventy-eight-seat plane in early 2004. Both companies' next-generation planes (the CRJ-900 and the ERJ-190/195) will seat more than one hundred people. These planes will list for about $30 million.[6]

Each improvement has inched regional jet manufacturers closer to established aircraft markets. Regional airlines could begin to displace traditional airlines on highly traveled routes (as we discuss later in this chapter). Next-generation planes could be good enough for traditional airlines to use for short routes.

If traditional customers begin migrating to smaller airplanes, Boeing and Airbus could be in big trouble. They would have to fight against their own skills and motivation to win this battle. Regional jet manufacturers make money on lower-priced planes. Regional jets largely exist within a freestanding value network, with almost no points of interaction with incumbent manufacturers. Regional jet manufacturers have developed the unique ability to build small and midrange planes profitably. Boeing's and Airbus's cost structures and capabilities support building large, sophisticated planes. Embraer's earnings per share increased about 30 percent annually between 1998 and 2003, while Boeing's earnings per share shrank at about a 5.5 percent annual clip over the same period.[7]

Can Boeing and Airbus respond to the onslaught of regional jet manufacturers such as Bombardier and Embraer? The theories of innovation suggest that Boeing and Airbus will continue to miss much of the industry's real growth and face increasing pressure unless they take proactive corrective action to adapt to the disruptive forces.

For instance, Boeing's hopes of staving off disruptive incursion would seem to rest on blocking advances from below through its 717.

The 717 is a modified version of McDonnell Douglas's MD-95 that seats about one hundred passengers. Not only will the 717 be more expensive than next-generation regional jets, but it also weighs significantly more, forcing operators to use more fuel. Because Boeing continues to base the plane on McDonnell Douglas's design, pilots have to receive specialized training to fly the 717. Even worse, it is too expensive for the lowest end of the market. Most analysts consider the 717 the worst plane in Boeing's lineup, making it an unlikely candidate for serious investment.

So what could Boeing do? The first option might be to set up an internal group to develop a small plane. Success would require overcoming two critical problems. First, Boeing might find it difficult to attract the right people to the project team. In addition, Boeing would have to fight its own cost structure lest it develop a "big little" plane—a small plane weighed down by Boeing's overhead and development process.[8] Unless Boeing manages the process incredibly well, the outcome seems clear. Boeing's natural reaction would be to withdraw from the low end of the market and focus on staying abreast of Airbus. Signs indicate that is indeed what Boeing is trying to do. Its latest big push is the 7E7. The 7E7 will fly farther and more efficiently than similar-sized planes. It is a classic sustaining innovation.[9]

As a second option, Boeing could try to address this problem by setting up a separate organization to develop a competing regional jet. Of course, at this point Boeing would be entering on Embraer's and Bombardier's sustaining trajectories. They are the incumbents. Boeing would be the entrant. The disruptive innovation theory would predict a low probability of success. The third option would be for Boeing to acquire a regional jet manufacturer and allow it to operate freely. This would be expensive, but would at least allow Boeing to catch some of the disruptive growth. Ongoing success from this pathway would require that Boeing keep this new acquisition as separate as possible from its core operations.

In this circumstance, the theories of innovation suggest betting on the attackers. They have hidden behind the shield of asymmetric motivation while creating asymmetric skills.

The high end of the market will not disappear overnight. Despite recent hardship, the flagship airlines continue to demand next-generation planes. However, unless Boeing and Airbus take corrective action, they face the prospect of spending the next decade focusing their growth investments on an increasingly expensive and inexorably dwindling pie.

INDUSTRY SUMMARY

Airplane Manufacturers

Signals of Change: Regional jet manufacturers have created growth in new markets and are now encroaching on the established market.

Competitive Battles: Incumbents' processes and values are liabilities, giving the potential edge to entrants.

Strategic Choices: Incumbents potentially could decide to drive their own disruptive growth.

Developments in Airlines: From Southwest to Air Taxis

Now we turn from the airplane makers to the commercial airlines. How can we make sense of the innovative business models circling around the airline industry? We touch on four topics in this section. First, we discuss how Southwest Airlines's disruptive growth based on point-to-point flights between second-tier airports could be reaching its limits. Second, we show how discount airlines can enter the industry but struggle to sustain profitable growth because the industry's economics make "flight" unattractive to incumbents. Third, we look at how regional airlines' decision to overlap with existing value networks created a business model that *looks* co-optable but could ultimately prove troubling. Finally, we show the decisions a hypothetical point-to-point air taxi company could make to launch a new-market disruption that could radically reshape the industry.

We start by looking at the entrenched incumbents in the United States—traditional national and international carriers such as American Airlines, Continental, Delta, Northwest, United Airlines, and US Airways.

Incumbent Overview and Tale of the Tape

By all accounts, the 1980s and 1990s should have been happy times for the aviation industry. The combination of an economic downturn and the tragic events of September 11, 2001, obscured what had been a two-decade-long boom in air travel. Between 1980 and 2000, the number of revenue-generating passengers boarding commercial planes grew to more than 600 million a year from about 300 million a year. The industry-wide average load factor (defined as the percent of available seats filled) increased almost linearly, growing from about 59 percent in 1980 to an all-time high of roughly 72 percent in 2000.[10] Major airlines each had strong market shares in their key markets. Such positive data would suggest an industry full of profitable growing companies.

But during this period, the largest airlines struggled to be consistently profitable. Industry-wide net income was positive only eleven times between 1980 and 2000.[11] The industry lost money nine times—in the years from 1981 to 1983, in 1986, and in the years from 1990 to 1994. The six largest U.S. airlines (American, Continental, Delta, Northwest, United, and US Airways) had total revenues of more than $600 billion between 1992 and 2000. Their combined net income over that period was a miserable $13.5 billion.[12] In contrast, Microsoft had net income of $9 billion in 2000 alone. And then when things got bad, they got really bad. In 2002 and 2003, goliath after goliath (including United and US Airways) filed for Chapter 11 bankruptcy protection. Airlines seem to be in a crummy business in which even the best firms have to fight hard just to survive.[13]

Indeed, becoming a successful commercial carrier is extremely difficult. Entering the industry requires a massive amount of capital because large planes can cost hundreds of millions of dollars. Expensive labor and fuel drive up operating costs. Safety standards are rigorously high. Operating a commercial aviation company requires managing a complex value network with thousands of employees, including highly trained pilots, and an unbelievably complex logistics system. The way airlines cope with this complexity results in their processes and values, which ultimately create opportunities for new entrants.

Airlines have invested millions building their hub-and-spoke systems. Each carrier has a hub in a large city in a region that connects to many smaller spokes. The system allows them to increase load factors, spread out capital costs, and ensure that passengers have multiple ways to get to a destination.

Airlines created the system to cater to the needs of their most important and most demanding customers—business travelers. Business travelers demand the ability to make a convenient round-trip to just about anywhere in a day. They tend to be price insensitive, since they do not typically bear the cost of the ticket out of their own pockets. They are willing to pay steep price premiums to retain travel flexibility and have their needs met. Although business travel made up only about 10 percent of all miles in 2001, it made up more than one-third of total revenue.[14]

So how did incumbents size up to the tale of the tape in 2004? Their resources were obvious—their planes, their people, their hub-and-spoke networks, their customers, and their established brands. Their processes included managing far-flung networks, training pilots and crew, running complicated pricing plans, managing loyalty programs, and so on. Their values focused on the most demanding customer, the business traveler. There is a difference between the airlines' situation and that of other vulnerable-to-disruption industry leaders, however. Leading firms in most industries—such as steel and department stores—actually improved their profitability as they fled up-market and ceded the most price-sensitive customers. The major airlines can't do that. The high fixed-cost nature of the business means airlines cannot ignore or stop serving their least demanding customers—budget-conscious leisure travelers.

These skills and motivations allow airlines to survive in a turbulent industry. Paradoxically, they also contribute to the airlines' problems. They present opportunities for new companies to enter the industry. One of the first companies to enter behind the shield of asymmetric motivation was Southwest Airlines, which grew as a hybrid new-market/low-end disruptor.

Southwest: Disruptive Growth Through Replication with Limits

Southwest built two shields that made incumbent response difficult: its point-to-point route structure and its low-cost business model. In the 1960s, three Texan entrepreneurs formed an airline called Southwest to travel between Houston, Dallas, and San Antonio.[15] Most people who traveled between these cities had to travel by car or bus. Flights were few and were expensive. Southwest entered by competing on these routes against nonconsumption—seeking to make it so cheap and simple to fly that people would fly instead of drive. In addition, Southwest's founders made a critical decision to fly out of Houston Hobby and Dallas Love Field airports instead of Houston Intercontinental and Dallas–Forth Worth airports to save costs. The second-tier airports were delighted to have increased business and, because they had fewer flights, allowed Southwest to load and unload passengers quickly, saving on expensive airport fees.

From this humble beginning, Southwest continued to expand to other cities. Its business model continued to focus on affordable travel between second-tier locations such as Baltimore, Maryland; Manchester, New Hampshire; and Oakland, California. Thirty years later, it was the most profitable airline in the United States and, as of March 2004, had a market capitalization greater than American Airlines, Delta, Northwest, and British Airways combined.

Southwest's offering historically could be classified as inferior. You couldn't get everywhere via Southwest. It flew a limited number of routes and primarily offered point-to-point travel. It didn't provide free food. You did not get an assigned seat. But its consumers were delighted with this relatively simple service offering because the alternative for many was not to fly at all.

Southwest's approach of only using airports in smaller cities provided its first shield of asymmetric motivation. Major airlines were not interested in playing a point-to-point game, especially out of smaller cities. Southwest's business model, which allowed it to earn attractive returns at low price points, provided its second shield. How could Southwest do this? It used a standard plane, saving on maintenance and training expenses. Its no-frills offering saved operating expenses. Its speed at loading and unloading planes saved airport fees and enhanced aircraft utilization. Chapter 7 of *The Innovator's Solution* notes how capabilities to succeed originate in resources, and then migrate to processes, values, and culture. Southwest's critical strengths were its processes and business model that supported its position as a low-cost provider.

As of 2004, most of Southwest's growth had not come via the traditional formula of other disruptive companies. Instead of marching upmarket to take customers away from incumbents, Southwest replicated its business model by offering flights on new routes. It continued to face little competition on its routes because it stayed in its freestanding value network and fought the temptation to invade the mainstream hub-and-spoke system. Because there were a number of untapped, underutilized airports in the United States capable of handling large planes, this expansion model provided years of profitable growth.

But Southwest's strategy may have a limit. When it expands to all available second-tier airports, it must move up by offering more hub-and-spoke–like service, stop growing, or create a new disruptive growth market. In late 2003, signs appeared suggesting that Southwest planned to try to expand by directly taking on the major airlines. For example, it launched a head-on assault against US Airways when it announced that it planned to start flying in and out of Philadelphia International Airport by May 2004. Southwest's incursion is sure to lead to a response by US Airways. The critical question will be whether Southwest has so well honed its unique processes that it can win the head-to-head battle against a weakened competitor. It needs to be careful, however. There is a lesson to be learned from other discount airlines.

Discount Airlines Enter but Motivate Fight, not Flight

In the previous section we noted that traditional airlines overshoot some customers who can be satisfied with just the basics. Over the last twenty-five years, numerous so-called discount airlines have tried to launch low-end disruptions to reach these customers. U.S. discount airlines such as People Express, Reno, Florida Air, New York Air, World, Presidential Airlines, Spirit, American Trans Air, AirTran Airways (formerly ValuJet), and JetBlue all created growth at the lowest end of the market at one point or another. The story is similar in Europe. Although leading European carriers such as British Airways and SwissAir struggled over the latter half of the 1990s, discount airlines such as RyanAir and easyJet grew consistently and profitably. But all discount airlines have a problem. They can take advantage of business model asymmetries to capture a chunk of the low end of the market. But as soon as they take too big of a bite, the major airlines' high fixed-cost structure motivates them to fight back to recapture the volume at the low end of the market. In other words, discounters cannot use asymmetries of motivation to march up-market and take increasing volume without incurring response. The majors are vulnerable to disruptive attack, but they cannot flee. They must stay at the low end and endure attack after attack. This is why it is such a consistently unprofitable business.

One of the highest-profile discount airlines in the United States in early 2004 was JetBlue, founded in 1999. Like other discount airlines, JetBlue started by targeting high-volume long-haul routes—in this case,

New York City to Los Angeles (flying from JFK to Long Beach, which is just outside of Los Angeles).

JetBlue developed a fundamentally lower-cost business model. It had nonunionized labor and operated a new, standardized fleet of aircraft (the Airbus A320) that allowed its flight crew to operate any plane in the fleet. Its reservation agents all worked out of their homes, using Internet-based technology to take reservations (see "Travel Agencies, the Internet, and Relativity") and issue paperless tickets. It deployed kiosks at airports for passengers to print their own boarding passes. Its on-board service consisted only of a limited selection of snacks (including its signature blue potato chips) and a limited beverage service.

These differences added up. JetBlue's operating cost per passenger mile in 2003 was about six cents, almost 20 percent lower than Southwest and about 50 percent lower than the average major carrier (see figure 6-1).[16] Despite its low costs, it offered passengers attractive amenities such as personal DIRECTV and relatively wide, leather seats. JetBlue's IPO roared out of the gates. It was profitable in 2002 and 2003. As of March 24, 2004, its market capitalization of $2.3 billion was about 50 percent higher than American Airlines.[17]

Discount airlines such as JetBlue have all the hallmarks of a low-end disruptive innovation. Their business model is lower cost. They offer a relatively simple, low-cost product to the market's least demanding cus-

Travel Agencies, the Internet, and Relativity

Existing airlines have implemented Internet-based reservations technology as a sustaining innovation—a way to reduce cost within their established business model. However, the Internet was extremely disruptive to full-service, face-to-face travel agents. The Internet enabled the creation of new business models that let people search through ticket databases that previously were only accessible to expert travel agencies. Trying to sell tickets on the Web just wouldn't make sense for a full-service travel agency that has to support high overhead cost. Expedia, Orbitz, Travelocity, and Priceline all capitalized on the Internet to create new growth by allowing each of us to become our own personal travel agent.

FIGURE 6-1

Airline Operating Costs

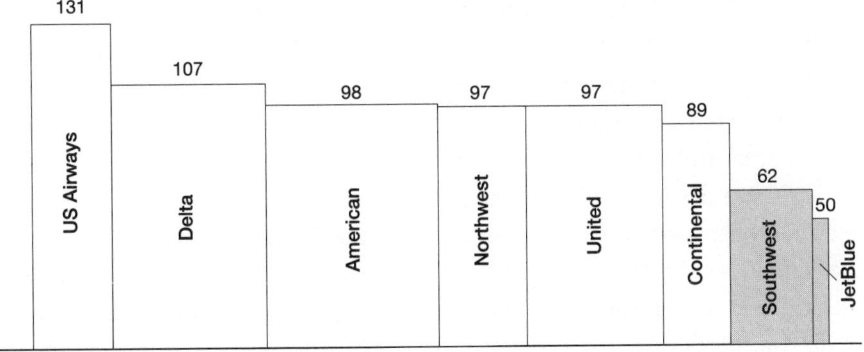

Operating Cost per Available Seat Mile, 2003

Indexed to average of major airlines = 100; width of bar indicates 2003 revenues

Source: Statistics compiled by authors based on figures gathered by the Bureau of Transportation; see <http://www.dot.gov/affairs/BTS2903.htm> (accessed 24 January 2004).

tomers. Leading airlines have tried to respond in two ways: by trying to create an airline within an airline and by aggressively fighting back. Let's assess each strategy.

Airlines Within Airlines

Almost every airline's effort to create a discount offering has been shut down—including US Airways's MetroJet, Continental's Continental Lite, United's West Coast Shuttle, and Delta's Delta Express. It is not surprising that analysts reacted with well-grounded skepticism when Delta announced at the end of 2002 that it planned to spend about $75 million to try again to build a discount airline within an airline called Song and United announced its hopes to move about 30 percent of its business to an internal discount offering code-named Starfish (later named Ted).[18]

Through the lenses of the theories of innovation, these struggles are not surprising. Internally managing a discount airline requires very different skills and a *very* different cost structure. An airline within an airline naturally leads to internal conflicts, ranging from angered unionized pilots to creeping expenses on the discount side, because of incremental choices that make it look increasingly like the traditional business.

Even when airlines successfully introduce a low-priced offering, they face severe pressure to shut it down whenever the parent gets sick and needs attention. The start-up expenses and low margins appear to be unnecessary resource drains in tough times. This experience mirrors Digital Equipment Corporation's (DEC) forays into personal computers. DEC spent $2 billion developing a personal computer. It entered the market four times, each time believing it *had* to be in PCs. But the profitability of its PCs could never match the profitability of its minicomputers. So each time it exited the market.

This would seem to be a clear circumstance in which incumbents' best bet would be to set up a legitimate spinout organization. If airlines did set up separate organizations, the thing to watch as the organizations grow and improve is whether their parent company truly gives them the freedom to cannibalize existing businesses. Efforts to create an airline within an airline without giving the new offering complete freedom are likely to ultimately be unsuccessful.

Aggressive Response

In the end, despite the problems major airlines have had in managing internal discount offerings, they have beaten back wave after wave of discount attackers by following their second response strategy: using their sophisticated dynamic pricing capability to, seat by seat, offer discount fares on regular flights. Remember, in chapter 2 we said the typical course of disruption involved incumbents' fleeing the low end and focusing on undershot customers up-market. Ironically, industry characteristics make "flight" next to impossible.

Why is this so? In short, relatively high fixed costs make it impossible to abandon the volume at the low end of the market. The consequences of abandoning large-volume markets would be too devastating.[19] Even the least profitable passengers contribute to the bottom line as long as they pay enough to cover the marginal cost (almost literally peanuts, or, these days, pretzels) of filling an extra seat. Furthermore, airlines have a limited ability to actually reach undershot customers. Conceivably, some customers demand higher on-time arrivals, more convenient flight times, and so on. But the airlines just don't have enough control over these factors to really create such a premium service. Nor do airlines have the freedom to develop an aircraft to support such a business model. They depend on the innovations that leading manufacturers spend decades developing. All of these factors motivate incumbents to fight, not flee.

Relatively low variable costs enable viable response strategies. Because the major airlines' marginal costs per extra passenger are much lower than the low-cost disruptors' average costs, the majors can cut price on individual routes and outlast the discounters. This hurts the incumbents and it hurts the discounters. For example, the day JetBlue flew its first flight from Boston, in January 2004, American Airlines announced that it would give a free ticket to any destination in the world to any passenger who flew two round-trips on overlapping routes. Delta similarly lowered its prices to drive JetBlue out of its Atlanta hub.

Discount Airlines Summary

Just because incumbents want to fight does not necessarily mean they will win the fight. They do have resources and extremely high motivation. The critical question will be whether JetBlue and other discounters have truly created business models to which incumbents cannot respond. Leather seats and DIRECTV will not be enough, because they are resource-based advantages, which are eminently imitable. To succeed, JetBlue and other discounters must develop unique processes and values that make incumbents unable and unwilling to respond. Our assessment suggests that a low-end approach does not create the requisite business model asymmetries. Therefore, as of 2004, the long-term prognosis for discount airlines—and the incumbents—is cloudy. A consistent battle for low-end customers means just about no one makes money.

Could discounters follow another approach? In the Southwest story, we noted one powerful advantage of Southwest's model was that it flew point-to-point flights from nonhub locations. Another tactic discounters can try is to take a new-market disruptive approach, avoiding incumbents altogether by going after routes incumbents don't fly. JetBlue's announcement in June 2003 that it planned to purchase a large number of regional jets could allow it to follow this path. JetBlue could enhance its chances of success if it uses those planes to fly point-to-point between secondary and tertiary cities, or if it uses the economics of regional jets to create minihubs to facilitate travel between smaller cities, rather than directly taking on the big hubs of major airlines. The next section looks at the regional jet operators who have grown in this exact fashion.

Regional Airlines Create Co-optable Business Models

Growing in tandem with regional jet manufacturers' disruptive ascendancy, regional operators teem with disruptive energy. They constantly seek solace in the arms of the leading airlines but might end up disrupting them.

Regional carriers came into prominence when they began buying regional jets in the early 1990s.[20] They used the first small regional jets to offer point-to-point travel to and from underserved areas. They are a low-cost alternative. Their planes cost less, require less expensive labor, and can fly into smaller, lower-fee airports. A regional flight breaks even if it is 50 percent full, whereas longer flights with bigger planes require a 63 percent load.[21] Regional airlines grew prodigiously through the late 1990s. The revenue passenger miles (RPMs) of the five largest publicly traded regional airlines (Atlantic Coast Airlines, ExpressJet, Mesa, Mesaba, and SkyWest) grew at a compound rate of 26 percent a year from 1997 to 2001. Over that same period, major airlines' RPMs grew at a measly rate of 1.6 percent a year.[22]

How did regional airlines achieve such strong growth? By complementing the hub-and-spoke system. In 2002, more than 60 percent of their flights fed on to the hub-and-spoke infrastructure of the large airlines.[23] This arrangement clearly placed regional airlines in an overlapping value network. It required creating a business model that complemented national airlines. Regional airlines chose to integrate their business tightly with the large airlines by sharing gates and using code-share agreements to allow a customer to buy a ticket from multiple operators at once. Some entered into fixed-fee arrangements that made them look very much like subsidiaries of leading airlines.

What happened when regional operators, using a business model that made sense to the national airlines, got big enough to matter? Just as the wireline telephone companies found it attractive to acquire and integrate wireless operators, some of the major airlines saw that they could easily use regional airlines to extend their hub-and-spoke systems. Delta purchased ComAir in early 2000. It also owns Atlantic Southeast. American owns American Eagle, which owns Executive, Flagship, Simmons, and Wings West Airlines. Before its initial public offering in 2002, ExpressJet (which flies under the name of Continental Express) was a wholly owned subsidiary of Continental. Continental retains a 53 percent ownership stake.

Co-option Might Still Have Consequences

These relationships have trouble written all over them. Major airlines think they have symbiotic, arms-length relationships with small regional airlines. But the disruptive potential of regional operators might be difficult to contain. Airlines might see the balance of power in the relationship suddenly shift.

Potential conflicts were invisible when regional jets could only fly short distances and carry limited numbers of passengers. As regional jet manufacturers increase the range and size of their planes, conflict could explode. Regional airlines are getting good enough to be a viable and cheaper way to fly longer, more profitable routes. Integrated airlines will face pressure to allow regional offerings to continue to grow.[24] Even those major airlines that have the contractual power to force their regional partners to stay small will feel pressure from independent regional airlines such as Atlantic Coast Airlines and Sky West.

Battles loom on the horizon. The factor that will continue to limit the disruptiveness of regional airlines is the overriding desire to complement rather than disrupt the existing hub-and-spoke system. Because regional airlines have felt that they need the major airlines to get passengers and to make connections, they have historically been unwilling to bite the hand that feeds them. Also, regional airlines that expand into routes that incumbents care about will face a similar problem to the discounters. Incumbents will be motivated to fight rather than flee.

Tough choices and growing battles may only exacerbate the industry's problems. Ultimately, it seems that large airlines will have no choice but to push for further co-option of regional airlines. Implementing regional jets with their associated operating cost savings as a sustaining innovation within their existing business model will prove just too tempting to ignore, especially if the industry continues to face intense profitability pressures.

Is there another way? Is there a structure that would allow a firm to craft a truly disruptive business model? As mentioned earlier in the chapter, a regional carrier could create a freestanding value network by setting up a mini hub-and-spoke system. The disruptive operator would cobble together a national network of secondary and tertiary airports. It would provide transport to anywhere in the United States without ever touching the traditional infrastructure. This freestanding value network would support an important asymmetry: Regional jet operators can create hubs at airports such as Raleigh-Durham, North Carolina, that would be too small to fit within a major airline's cost structure.

Even if a regional operator bent on disruption took all the right actions, it might find the disruptive way difficult. In some situations, it is legitimately difficult to create a freestanding value network. Only a limited number of airports can handle even relatively small regional jets.

But what about the thousands of small airfields that could support even smaller planes? More than 98 percent of Americans live within twenty miles of a public-use airfield.[25] As of early 2004, only corporations, aviation aficionados, and well-heeled fractional jet owners used these airports. But future generations of low-cost small aircraft could enable the creation of a point-to-point air taxi industry.

This chapter's final section points to the strategic choices a hypothetical Air Taxi Company (creatively called ATC) could make that would give it the best chance of reshaping the aviation industry.

Choices Emerging Air Taxi Providers Could Make to Build and Harness Asymmetries

As the cost of small aircraft declines, harried wealthy businesspeople are increasingly turning to either direct or fractional ownership of small planes to avoid dealing with the everyday hassles of flying. As of 2004, these point-to-point solutions were only available to those wealthy enough to afford flying on multimillion-dollar personal jets. But as jet prices decline, or if a company could develop a reliable plane that was inexpensive to purchase and operate (see "Disruptive Planes"), ATC could bring point-to-point travel to the masses.

ATC would face two critical choices: selecting a target market and deciding how to interact with existing providers. The models of disruption suggest the approach that could have the greatest impact would be to target nonconsumers by building a freestanding value network that facilitates incubating a business model capable of industry-wide disruption.

The first big issue ATC would face would be choosing its initial target market. One alternative would be to cherry-pick the best, most well-traveled routes, such as New York to Washington, D.C., or San Francisco to Los Angeles. Although short-term growth would be possible, this approach would surely raise the ire of incumbents who have a lot to

Disruptive Planes

In early 2004, New Mexico–based start-up Eclipse Aviation was working to develop a plane that could radically change the airline industry. Founder Vern Raburn tried to bring the principles he learned as the nineteenth employee hired at Microsoft to the aviation industry. With close to $300 million in funding from private investors (including Bill Gates), Eclipse developed a six-seat microplane known as the Eclipse 500. It can fly about 1,000 nautical miles (roughly the distance from Boston to Miami) at a top speed of 400 miles per hour. It will sell for about $1 million and have operating costs of about $0.75 a mile. That is about one-third of the price of its nearest competitors, Cessna Aircraft's citation CJ-1 (which costs about $1.50 a mile to fly) and less than 10 percent of the price of General Dynamics's Gulfstream. The Eclipse 500 needs about 2,000 feet of runway space to take off and 2,500 feet of runway space to land. More than 98 percent of Americans live within twenty miles of an airfield that could theoretically support Eclipse 500 jets. Eclipse plans to introduce the plane in 2006. At the end of 2003, it had more than 2,000 orders.

As of the writing of this book, the company still had to clear numerous technological hurdles and still needed to get government approval. But Eclipse could create tremendous disruptive growth. It has said it wants to be the Ford of aviation, bringing airplane ownership to the nonconsuming masses. Of course, Eclipse could also be a straightforward low-end disruption by targeting the most overshot purchasers of corporate jets. Regardless, it is worth watching.

lose. Our theories tell us to bet on companies that try to beat the easiest competitor: nonconsumption. ATC could target corporations or small firms that lack the financial resources to justify owning a corporate jet but have manufacturing plants, suppliers, or partners in obscure locations that are difficult to reach using traditional airlines. These nonconsumers would welcome a service that allowed them to visit their factories and suppliers with greater frequency. Another possibility would be for ATC to provide service in areas where there are no airports capable of handling even regional jets. There are many such areas in the United States and probably even a greater number in less developed countries.

Building several small airports (which amount to no more than a landing strip) and flying a few small planes might prove more cost-effective than building an entire transportation infrastructure. Industry players might ignore developments in these fringe markets. From ATC's perspective, that's a good thing.

What preparation regimen could ATC follow to maximize its chances of finding the right target market? Rather than purchasing hundreds of planes and trying to build a national network overnight, following an emergent strategy would be critical.[26] So instead of "build it and they will come," ATC would build to demand.

What schools of experience would be helpful here? A CEO who had used emergent strategy processes to create a new-market disruption in another service business would be ideal. It wouldn't necessarily be good news if ATC picked a well-respected senior manager from a leading airline to be its CEO. That CEO's schools of experience taught him or her how to run a very different business. In fact, this sort of CEO would quite naturally look to the markets he or she knows best, the large and measurable ones.

And what sorts of investors would enhance ATC's chances of success? Investors need to be patient for growth, but impatient for profit, to help the right strategy emerge faster. Capital from investors who desperately need the venture to get very big, very fast could be a death sentence. Trying to grow quickly would push ATC toward meeting the needs of the very demanding customers who matter a great deal to powerful incumbents.

The other critical choice ATC would face would be its selection of a value network. The easy choice again would be to create a feeder model that overlaps with the existing airline network and in return drives passengers to its planes. But as chapter 3 suggested, entering an overlapping value network would ultimately dampen ATC's long-term value-creation potential. A freestanding value network limits incumbents' ability to coopt the innovation. It would let ATC reach more nonconsumers, giving it enough time to develop unique skills to which incumbents could not respond.

ATC would of course have to wrestle with other problems. Creating a robust air taxi infrastructure could be very complicated. For example, when we fly today, we take for granted that the arriving airport will have a well-developed taxi and rental car infrastructure that helps us

INDUSTRY SUMMARY

Commercial Airlines

Signals of Change: Discount airlines are creating low-end disruptive growth; regional airlines are creating growth in a new context.

Competitive Battles: Incumbents tend to be motivated to fight discounters; overlapping value networks lead to co-option of regional airlines.

Strategic Choices: Key decisions to watch are target routes and value network design; freestanding value networks and a new route structure could encourage disruption.

---

reach our final destination. Although it might be a constraint, it also would represent an opportunity for a firm such as Enterprise Rent-A-Car to launch its own new-market disruption.

We think these are solvable problems. Getting the target market and value network right could position ATC to be one of the great growth companies of the twenty-first century.

In the short term, theory suggests the airline industry's future will look somewhat similar to its past. But long-term trends and forces could ultimately lead to a very different industry structure.

On the manufacturing side, Boeing and Airbus will continue to slug it out for control of the high end of the market. Their own head-to-head battle is likely to distract them from the threat they need to worry about the most. Regional jet manufacturers will continue to soar up-market. One day Boeing and Airbus could wake up and realize they are fighting over a declining market. To put it starkly, Boeing has two choices: to buy Embraer or Bombardier or to disrupt them by creating a completely new market or a viable low-end plane.

Leading commercial airlines face threats on many fronts. Discount airlines continually enter and nip at their heels. Because there is not asymmetric motivation, disruptive discounters are unlikely to transform the industry. But they will be a perpetual nuisance to incumbents who have no up-market customers to target. Discounters that find ways to lessen the odds of incumbent retaliation have the best long-term prospects.

Regional airlines drove a good portion of the industry's growth in the 1990s. Although national operators seemingly co-opted regional jet operators, that relationship has conflict written all over it. Regional airlines' improvement will add further pressure to incumbents. But because even independent regional airlines live in overlapping value networks, the likely result will be fierce battles on all sides without a radical industry reshaping.

Troubled skies loom. It is very possible that in the future no one will be able to make any money. So what can we watch for that has legitimate breakout potential? Leading incumbents can turn the forces of disruption to their advantage by creating spinout organizations or developing internal capabilities. Discounters and independent regional airlines can make bold choices to execute new-market disruption by developing freestanding value networks. Emerging point-to-point air taxis can create new value networks that reach nonconsumers. Any of these developments could radically reshape the entire industry.

The aviation chapter has five general lessons:

- *Skills and motivation define companies' strengths and weaknesses, creating opportunities for new sorts of companies and determining companies' likely response to disruptive growth.*
- *Companies can create growth without incumbent response by leveraging asymmetric motivation (such as Embraer, Bombardier, Southwest).*
- *An industry's context can affect how companies react to disruptive threats (such as airlines not being able to flee the low end, forcing a fight).*
- *Companies with disruptive potential can choose co-optable business models by situating themselves in overlapping value networks (such as regional jets).*
- *A company's choice of initial target market and value network can increase its disruptive potential.*

Notes

1. Different generations of planes (for example, the 737-600, -700, and so on) could be considered incremental sustaining innovations, although they still involve millions of dollars in development.

2. The U.S. Commerce Department reported in 1990 that France, Germany, Spain, and the United Kingdom directly provided about $13 billion to Airbus. A consulting firm estimated the total subsidization as of 1990 to be about $26 million including forsaken interest costs. The European companies responded that the United States provided up to $22 billion in implicit subsidies to McDonnell Douglas and Boeing through defense contracts. The European governments agreed to curtail the subsidies in 1992, but the United States contended that governments provided another $4 billion in subsidies for the A380. See "Government May Take Action Under GATT Over Airbus Subsidies," *Flight International*, 22 May 1991; Brian Coleman, "Airline Industry: EC Report Fires Back at U.S. in Jet-Firm Subsidy Dispute," *Wall Street Journal Europe*, 5 December 1991; Adrian Cox, "U.S. to Challenge Airbus Subsidies; EU Loan to Boeing Rival Under Dispute," *Bloomberg News*, 29 October 2002.

3. Our analysis focuses on the two leading regional jet manufacturers. Other players include Fairchild Dornier, which fell on hard times in 2002, and Avions de Transport Regional (ATR).

4. Embraer's roots were in military applications. In fact, Brazil and Canada got into a huge trade dispute over the issue of subsidies. Canada complained that Brazil implicitly gave Embraer illegal export subsidies through its defense relationships.

5. Regional jets could be viewed as a radical sustaining innovation vis-à-vis turboprop planes. Turboprop planes had limitations. They were noisy and uncomfortable. Interestingly, this appears to be a case in which at least some incumbents (turboprop manufacturers) failed to master a sustaining innovation. The same thing happened when jet engines supplanted piston-propeller engines in large commercial aircraft. The airline industry is full of interesting developments such as this one that warrant much deeper investigation.

6. Figures from Embraer and Bombardier annual reports, company Web sites (<http://www.embraer.com> [accessed 27 September 2002] and <http://www.bombardier.com> [accessed 4 February 2003]), various issues of *Aircraft Value News*, and analyst reports (see, for instance, "Regional Jet Rentals Continue to Ease," *Aircraft Value News*, 6 October 2003; Carlos Albano, Christian Flemming, and Luciana Zonzini, "Embraer," Unibanco/Stern, Stewart and Co., 25 July 2002).

7. Charles Schwab online research by authors, <http://www.schwab.com> (accessed 25 February 2004).

8. In personal conversations with Clayton Christensen, Harvard Business School dean Kim Clark observed the same phenomenon in the automotive industry.

When Detroit automakers responded to the incursion of Japanese automakers, they produced small cars that were relatively expensive because the automakers tried to cram the functionality of a large car into a small car.

9. For a remarkable discussion of Boeing's tough choices, see J. Lynn Lunsford, "Navigating Change: Boeing, Losing Ground to Airbus, Faces Key Choice—Amid Downturn, Firm Ponders Spending on Innovation or Diversifying Defensively—Debate Over All-New Plane," *Wall Street Journal*, 21 April 2003; and Stanley Holmes, "Boeing: What Really Happened," *BusinessWeek*, 15 December 2003.

10. Numbers represent an average of U.S. Bureau of Transportation statistics and numbers reported individually by major carriers.

11. Industry figures from the U.S. Bureau of Transportation Statistics, available from <http://www.bts.gov/oai/indicators/top.html> (accessed 23 September 2002).

12. Compiled by authors based on figures from CompuStat PC Plus, Standard & Poor: Englewood, CO (accessed 12 March 2003).

13. A five-forces analysis would suggest the industry is structurally unprofitable because of relatively low barriers to entry (due to low switching costs), high concentration of supplier power (although that will diminish as regional jets continue to move up-market), and high degrees of competitive rivalry among firms (one of the major drivers of this rivalry is high exit barriers). However, that answer is at least somewhat counterintuitive given the industry's steep scale economics and inflexible route structures.

14. Jim Corridore, "Airlines," S&P Industry Survey, <http://www.S&P.com> (accessed 24 September 2002).

15. For an interesting, entertaining, and informative study of Southwest, see Jackie Freiberg and Kevin Freiberg, *Nuts! Southwest Airlines' Crazy Recipe for Business and Personal Success* (New York: Broadway Books, 1996).

16. Statistics compiled by authors based on figures gathered by the U.S. Bureau of Transportation Statistics. See "BTS Releases Third Quarter 2003 Airline Financial Data; Regional Passenger Airlines Report Highest Rate of Domestic Profit," 23 December 2003, <http://www.dot.gov/affairs/BTS2903.htm> (accessed 24 January 2004).

17. Figures from Yahoo! Finance, <http://finance.yahoo.com> (accessed 24 March 2004). The market capitalization of American's parent company, AMR, was $1.69 billion.

18. Sonoko Setaishi, "Delta, United Low-Fare Units Greeted With Skepticism," *Dow Jones News Service*, 20 February 2003.

19. The need to maintain margins is one reason why companies that move up-market tend not to do it by taking an existing product up-market. That would confuse customers and damage built-up existing brand equity. They tend to introduce a new product or service or use a separate brand. When Toyota moved up-market, it didn't try to take its lowest-performing cars and introduce them to more demanding customers. It introduced completely new models, such as the

Camry, Lexus, and so on. Similarly, Coca-Cola simply could not take its Coke brand to customers who demanded fruit drinks, flavored teas, and branded water. It could only go after those markets by launching or acquiring new brands, such as Dasani or Fruitopia.

20. Of course small airlines before that flew turboprop planes. To smaller operators, the regional jet was a highly sustaining innovation, which is one of the reasons regional jet manufacturers found such success. They provided the fuel that allowed smaller regional operators to move up-market.

21. Corridore, "Airlines."

22. Figures from various analyst reports. See particularly: Michael J. Linenberg and Sandra Fleming, "Airline Industry Quarterly Review," Merrill Lynch, August 2002; Susan Donofrio and Allison Poliniak, "Airlines 101: An Introduction to the U.S. Airline Industry," Deutsche Bank Securities Inc., August 2002; William J. Greene, "Regional Airlines: Stable Growth in a Volatile Industry," Morgan Stanley, 1 March 2002. Reports available from Investext. Additional information available from the U.S. Bureau of Transportation Statistics, <http://transtats.bts.gov> (accessed 24 March 2004).

23. Corridore, "Airlines."

24. Large airlines' union contracts typically forbid this type of cannibalization even if it would be good for the overall business. As of 2003, only Continental Airlines, Alaska Air, America West, and Southwest did not have so-called scope clauses that limited their use of regional airlines. The financial pressure leading airlines faced in 2003 led many of them to revisit scope clauses. Major airlines ultimately might have no choice but to let the disruptors expand.

25. John Croft, "Small Airports: To Be or Not To Be? NASA's Vision of a Future Full of Aerial Taxis Depends on Airports That Are Under Siege and Rapidly Disappearing," *Aviation Week & Space Technology*, 15 April 2002. For more information on the potential of air taxis, see James Fallows, *Free Flight: From Airline Hell to a New Age of Travel* (New York: Public Affairs, 2001).

26. Of course, if signals emerge that ATC has stumbled on the right business model, it should seek aggressive growth.

CHAPTER SEVEN

WHITHER MOORE'S LAW?

The Future of Semiconductors

Why did entrants, not incumbent firms, create the semiconductor industry? What problems have companies had to solve? Why did the value chain organize the way it did to solve those problems? Are there signs of overshooting? What implications would overshooting have on the value chain's structure? Would this be good news or bad news for incumbent leaders such as Intel? What sort of firm could capture growth if circumstances change? What portions of the industry will grow profitably in the future, which will languish and slide toward commoditization, and why?

Practically synonymous with high technology, the very term *semiconductor* conjures up images of massive manufacturing plants filled with workers clad in futuristic white "bunny suits."[1] The semiconductor manufacturing process is so sophisticated that plants must be immaculate lest an unobservable particle of dust destroy a chip. World-renowned firms such as Intel, Motorola, Toshiba, Texas Instruments, Samsung, and Taiwan Semiconductor populate the industry.

Everyone knows semiconductors power personal computers. But these days, semiconductors are in just about everything. Automobiles have upwards of fifty microprocessors controlling everything from the braking system to the engine. Consumer devices such as cellular phones and DVD players rely on semiconductors.

This chapter uses our approach to analyze the past, present, and future of the semiconductor industry. We begin by showing how the theories of innovation help to explain the industry's past. The industry's history traces back to a new-market disruptive innovation in the 1940s: the transistor. Because the industry is complex and unfamiliar to many readers, we will go into some detail about its structure today. But we won't ignore theory—the VCE and disruption theories help to explain why the industry is structured the way it is.

After we set the stage, we will then look for signals of change. Although the previous chapters presented numerous opportunities to explore growth in new markets and new contexts, the analysis in this chapter focuses on a single, provocative question: Are there signs that microprocessor manufacturers are overshooting the performance their customers can utilize?

We suggest the answer is yes. For decades, the industry has been obsessively pursuing something called Moore's Law. The prediction, posited by industry pioneer Gordon Moore more than thirty-five years ago, has been that by shrinking the width of conductor lines on a chip, the number of transistors on a given area of silicon wafer will double every eighteen to twenty-four months without a corresponding increase in cost.[2] Performance doubles; costs remain the same. By 2004, signs had emerged suggesting that companies pursuing Moore's Law were in danger of overshooting the needs of customers in less demanding tiers of the market.

Overshooting has profound implications. New companies could compete based on convenience and customization. Meeting these goals would require a reconfiguration of the industry's value chain. Success in the old world would not guarantee success in the new world.

What semiconductor manufacturers have achieved in meeting Moore's Law over the past thirty-five years is remarkable. But the very things that fueled the industry's success have created a circumstance in which entrants could reshape the industry. Viewing the semiconductor industry through the lenses of the theories of innovation suggests a high probability that the future will look very different from the past.

Setting the Context: Organizing to Meet Moore's Law

The semiconductor industry grew out of a disruptive innovation that took root in surprising niche markets. Subsequent sustaining innovations focused on improving speed and capacity. To improve what was not good enough, certain pieces of the value chain had to be optimized and interdependent. Conservation of integration implies that modular and configurable pieces had to surround the interdependent, optimized pieces.

Industry History and Overview: From Disruption to Digital Signal Processors

Researchers at Bell Labs developed the first transistor in 1947. A transistor is like a light switch: It can essentially be on or off. A single transistor can only do one task. Combining transistors and giving each one a specific task allows for the miracles of science that we take for granted today, such as televisions, telephones, and computers. When Bell Labs announced the transistor's invention in 1948, the result was a collective yawn. The *New York Times* carried the news at the tail end of a column on page forty-six titled "The News of Radio." Deemed more important was a notice that "'Waltz Time' will be heard for a full hour for three successive Friday evenings at 9 o'clock over NBC."[3]

The first transistors weren't very good compared to the vacuum tubes that powered electronic devices such as televisions and tabletop radios. Although transistors were a lot smaller and more rugged, they could not handle the power that manufacturers of products that relied on vacuum tubes required. Also, vacuum tube–based products had spawned a sprawling value network of retailers, distributors, and repair shops. All of these value network participants made money based on the vacuum tube's unique properties.[4]

But the transistor's unique properties—its small size and lower power consumption—made it a perfect tool to assist the hearing impaired. In 1952, Sonotone replaced one of the three vacuum tubes that powered its hearing aid with a transistor manufactured by a small Jersey City, New Jersey, company called Germanium Products Corp.[5]

The next big commercial market for transistors was pocket radios. Again, the unique properties of the transistor made it a perfect fit for the pocket radio. The first pocket radios weren't very good. They sounded tinny. But teenagers had a job to get done. They wanted to listen to rock music away from the disapproving ears of their parents.[6] Sony's transistor radio nailed this job and created an entirely new market by competing against nonconsumption. Even the poorest of sound qualities delighted teenagers, because for most of them, the alternative was to have no radio at all.

Transistors, of course, got better and ultimately heralded the end of the vacuum tube. What happened to the vacuum tube manufacturers? They saw the transistor from the beginning. They tried to co-opt the innovation and invested the equivalent of hundreds of millions of today's dollars to improve the transistor. Manufacturers had to make large investments, because the first transistors were not nearly good enough to serve their core markets—tabletop radios and floor-standing televisions. The established companies launched big development efforts to fix the transistor's current-carrying abilities to make it good enough for their core markets. They tried to cram the transistor into their existing products, but they failed. None of the giants of the vacuum tube era—such as Westinghouse, RCA, Zenith, General Radio, GE, and Raytheon—became big players in the emerging semiconductor industry.

Sustaining innovations following the introduction of the transistor led to the creation of the modern semiconductor industry. In 1958, a team of engineers found a way to put two transistors on a silicon crystal, creating the first integrated circuit. The integrated circuit evolved into products such as memory chips and microprocessors. The industry grew at a compound annual rate of almost 14 percent a year from 1960 to 2002.[7]

Today, the semiconductor industry consists of three distinct types of companies that participate in chip design and manufacturing. The first type of company, known as integrated device manufacturers (IDMs), does both design and manufacturing. The second type of company, known as foundries, focuses solely on manufacturing. Making a semiconductor is the one of the most complicated manufacturing processes the world has ever known, requiring more than four hundred discrete steps. Even ancillary processes are incredibly complicated. Intel's Web site describes the forty-three steps its workers follow when putting on their bunny suits.[8] As manufacturing facilities, known as wafer fabrication facilities or "fabs," grew more expensive and complicated, specialist manufacturers that did no design work emerged. The leading foundry is Taiwan Semiconductor Manufacturing Company (TSMC). TSMC's revenues increased more than 33 percent a year from 1993 to 2002, with 2002 revenues of almost $4.7 billion.[9]

Finally, the third type of company only designs chips, leaving manufacturing to the foundries. These companies, such as Xilinx and Tensilica, are known as "fabless" companies.

Two other types of ancillary players support the manufacturing process. Companies such as Applied Materials and Tokyo Electron make manufacturing equipment. Another group of companies makes equipment to test semiconductor products. Leading test equipment companies include Teradyne and Agilent.

Semiconductor companies sell their products to manufacturers of devices ranging from personal computers to personal digital assistants to automobiles. In 2001, the three largest end markets for semiconductors were computers (about 55 percent of the market), communications devices (about 25 percent), and consumer goods (about 15 percent).

Microprocessors, a $40 billion market in 2001, provide the "brains" of most computers. Intel dominates this market with more than 80 percent market share. Demanding server and workstation manufacturers use the most powerful microprocessors. Less demanding personal computer manufacturers use somewhat less powerful microprocessors.[10]

Manufacturing Process and the Conservation of Integration

Historical demands for higher-performing chips necessitated creating a manufacturing process that was optimized in the right places and customizable in the right places. The VCE theory and the law of conservation of integration explain why and where optimization occurred and why and where customization occurred.

Historically, the speed of the microprocessor has not been good enough. The overwhelming focus of the industry's value chain has been to optimize the microprocessor's performance.

What do value chain participants do to meet this goal? Besides the materials used in making the chips, the major factor that determines the speed of a device is the distance electrons must travel. As more of the device's functions can be packed onto a single silicon chip, these distances collapse, making the devices faster. The thinner the width of the conductor lines that comprise the circuit, the more functions that can be crowded on the chip. Therefore, the manufacturing equipment has to be specialized. Each individual production step has to strain to make the thinnest possible features and lines. Pushing these limits involves high degrees of creativity and experimentation.[11] The output, when it works, is as good as it can possibly be. However, because companies are working on the frontier of what is possible, the failure rate of each individual process step is quite high and unpredictable. And because companies don't want to waste time working on semifinished output that they already know is defective, after each process step they must put the interim output in a buffer of inventory, test it, and move it back to the buffer before finally moving it to the next step. The buffer is necessary for two reasons. First, companies never know exactly how much is going to make it through the testing stage. Second, this process just can't be run as an assembly line. Optimization requires different setup times and asynchronous cycle times for each process step. Having buffers and work in process (WIP) is the only way to cope with this complexity.

Operations management experts consider this a "batch" process. Manufacturers cluster machines that perform a particular function together into a "bay." They configure each bay to optimize the equipment's performance and utilization. Large batches of wafers are processed through a machine in one bay and then transported in a batch to machines in the next bay, where the wafers are tested and the next process step is performed on the entire batch, and so on. The inventory buffers the interface between bays, alleviating the need to synchronize the flow of wafers across the machines at the same pace as the sequence of manufacturing steps is performed. With the asynchronous cycle times and the need to continually test and store, the production process is expensive and relatively slow. Just like shoppers moving step by step up the long checkout line in a crowded supermarket, wafers move through these inventory-filled plants slowly, unable to move ahead until the wafer before them has moved one step farther. In 2004, it could take up to three months for a batch of wafers to work its way through a fab.[12]

The VCE theory and the related concept of conservation of integration explain what this value chain *must* look like. Remember, conservation of integration holds that whatever needs to be optimized needs to be surrounded by something that is configurable. As shown in figure 7-1, the architecture of the machines that power each process step must be interdependent and proprietary in order to make the finest feature sizes possible. To support this optimization, buffers of WIP inventory serve as modular interfaces between process stages. In other words, each process *step* is optimized. But the linkages between the steps are modular. This modular process creates a processor, which within itself is an interdependent, optimized architecture. End devices such as computers are modular and configured around the optimized microprocessor. Inte-

FIGURE 7-1

Conservation of Integration in Semiconductors

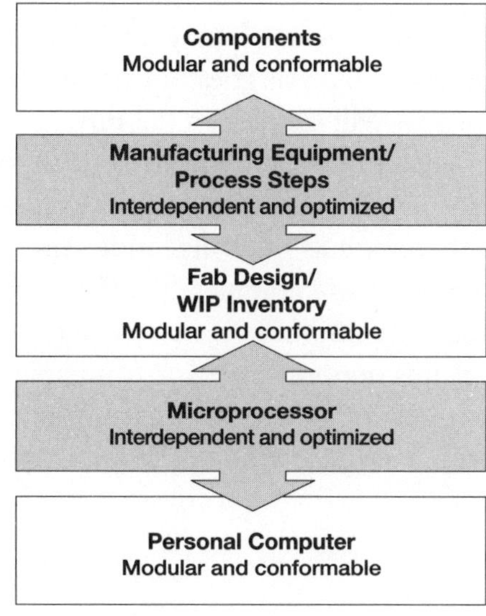

gration is conserved at alternate steps throughout the process, each step mediated by something modular.

Signals of Change Point to Overshot Customers

Historically, this approach worked brilliantly. Over the past thirty-five years, scientists, designers, and manufacturers have consistently come up with unanticipated ways to continue to make Moore's Law happen. Product designers improved chip performance like clockwork. The improved performance delighted customers. Users richly rewarded companies that produced chips with faster processing speeds. Increasingly powerful semiconductors enabled applications we could only dream of a generation ago.

The result has been unprecedented economic gains. Even in the midst of a horrific downturn in 2001, the industry chalked up revenues of almost $140 billion. Experts attribute roughly one-sixth of the increase in productivity in the United States in the latter half of the 1990s to improvements in the semiconductor industry.[13]

But that was the past. What about the future? Most analysis of this industry begins and ends with a single question: Will companies continue to be able to follow the trajectory of improvement defined by Moore's Law? The implicit analysis goes: If the answer is yes, companies that succeeded in the past will succeed in the future; if the answer is no, companies that succeeded in the past will struggle in the future.

But what if performance continues to follow Moore's Law, and it doesn't matter? Are there signals of change indicating that success in the future will require competing in completely different ways?

Our analysis suggests companies have overshot some market tiers. This overshooting creates opportunities for companies to enter, offering convenience and customization. Meeting this new value proposition will require a fundamental reconfiguration of the industry's value chain.

Signs of Overshooting

Moore's Law will merit its primacy in the microprocessor business only so long as there are undershot customers demanding faster microprocessors, or there are unmet needs to integrate more and more functionality on a given area of silicon. In early 2004, there were strong signals that undershot customers continued to exist in the demanding tiers of the market. Although servers are reaching the point where they are more than good enough for most applications, emerging applications such as three-dimensional games, digital video file editing, and real-time conversion of speech to text will continue to tax even the fastest microprocessors. Designers of these products are likely to continue to be willing to pay substantial premiums for products that get them closer to meeting their needs. As a general rule, however, unit volumes in these most performance-demanding tiers of the market are a fraction of those in mainstream market tiers.

Understanding whether existing companies will be able to delight these undershot customers requires determining the practical life of Moore's Law. If history is any guide, motivated innovators will continue to do the seemingly impossible and find unanticipated ways to extend the life of Moore's Law. Although there is much consternation that at some point Moore's Law will run into intractable physical limits, the only thing we can predict for certain is that innovators will be motivated to figure out solutions.

But this does not address whether meeting Moore's Law will continue to be paramount to success. Everyone always *hopes* for the emergence of new, unimagined applications. But the weight of history suggests the unimagined often remains just that; ultimately ever more demanding applications will stop appearing or will emerge much more slowly than anticipated. But even if new, high-end applications emerge, rocketing toward the technological frontier almost always leaves customers behind. And it is in those overshot tiers that disruptions take root.

How can we tell if customers are overshot? One signal is customers not using all of a product's functionality. Can we see this? There are ever-growing populations of users who couldn't care less about increases in processing power. The vast majority of consumers use their computers for word processing and e-mail. For this majority, high-end microprocessors such as Intel's Itanium and Pentium 4 and AMD's Athlon are clearly overkill.[14] Windows XP runs just fine on a Pentium III microprocessor, which is roughly half as fast as the Pentium 4. This is a sign that customers may be overshot.

Another signal is the ability of specialist companies to gain traction in the industry. Remember, specialist companies can never match the pure functionality of integrated companies. They can move faster and offer more customized solutions, however. Can we see this? IDMs initially scoffed at the foundries because they could not believe that contract manufacturers could master such a complicated process. But by 2004, even the top IDMs outsourced at least some of their production to the foundries. The emergence of small fabless companies is another sign that customers may be overshot.

These signals indicate that by 2004 companies had begun to overshoot some customers. At one level, this assessment is not surprising. Meeting Moore's Law means processing power improves about 60 percent every year. But designers' ability to utilize transistors is only increasing at about 20 percent per year. Their new product design cycles and product design budgets—not to mention the pace at which our lives change—simply means that we cannot use transistors at the same pace that Moore's Law makes them available. Designers of less complex circuits are awash in transistors they cannot use. The National Technology Roadmap for Semiconductors, a document produced by the Semiconductor

Industry Association to assess the future technological requirements for the industry, noted this so-called design gap in the late 1990s.[15] When the chips weren't even close to good enough for the market's needs, this design gap was not a problem. But when chips become more than good enough for the market's needs and the ranks of overshot customers swell, the game can change.

Implications of Overshooting

Overshooting creates opportunities for firms to change the basis of competition. Instead of competing based on raw functionality, firms can compete based on the ability to give customers exactly what they want, when they want it. What overshot customers consider not good enough, in other words, flips from performance to convenience. This change means that those who buy chips will begin demanding exactly the features and functions they need and will grow increasingly unwilling to pay for features and functionality they *don't* need. To deliver these specialized, task-oriented chips, the entire industry will have to organize in a very different way. Change will be dramatic, beginning in lower tiers first, and then creeping up-market.

What does customization and convenience look like in this industry? Consider Tensilica. A Silicon-Valley start-up, Tensilica allows engineers to customize their own systems-on-a-chip on a Web site. Engineers can mix-and-match intellectual property blocks to create a customized product. Think of these intellectual property blocks as Legos. Some Legos are small and do very simple things, such as control a serial port interface. Some Legos are bigger and do things that are more complicated. By piecing together a unique group of intellectual property blocks, engineers can create a fully functioning processor that perfectly suits their particular product. Tensilica's products cannot match everything that a one-size-fits-all microprocessor can do. But engineers can customize and optimize performance for a particular end use, such as a handheld device optimized for the medical industry.

In the world of logic chips, a similar group of companies sacrifices functionality in the name of convenience. The leading provider of programmable logic chips is Xilinx.[16] Founded in 1984, Xilinx essentially packs a lot of functionality on a chip and leaves it up to users to decide what specific functionality they need. This necessitates wasting functionality. But it allows engineers to design chips that meet their specific demands. Xilinx has a strong position in networking equipment and consumer electronics segments. Its revenues reached more than $1 billion in 2003.

A generation ago, for customers to personalize or custom-configure their own processors would simply have been impossible. Highly trained experts took years to design chips. What facilitates customization? In chapter 1 we noted the ways in which the development of widely accepted rules allows users closer to the end consumer (in this case engineers at product companies) to do what historically required deep expertise. Design rules developed over the past twenty years to help companies move along their sustaining trajectories allow less-skilled people to produce good-enough products (see "Rules in Semiconductors").

Indeed, Tensilica ultimately aspires to make assembling blocks of design into a piece of silicon so straightforward that even software engineers could create processors tailored to specific software applications.

Without overshooting, having nonexperts use rules to create new products would be impossible, because striving to reach the technological frontier entails breaking rules. But in an overshot world, engineers within companies can follow the rules, waste some functionality, and design customized chips good enough for their purposes.

Of course, someone still needs to produce customized chips. Next, we will discuss how changes in the basis of competition affect the manufacturing process. We will show that when what is not good enough changes, the entire value chain has to change as well.

Implications for Manufacturers: From Integrated Process Steps to Integrated Process

The increasing number of fabless designers has been a boon for foundries, historically viewed as an unexciting portion of the industry. But meeting the new needs has important implications for the manufacturing process, the companies that support the manufacturing process, and where the money can be made. The principle of conservation of integration helps to show how integration has to shift throughout the value chain to optimize what would now be not good enough.

Rules in Semiconductors

Early semiconductor design was very complicated. Only highly trained experts had the requisite skills. In an interview in the *Harvard Business Review*, Cypress Semiconductors CEO T. J. Rodgers recalled the process of chip design in the 1970s: "The biggest bottleneck was getting access to the central Burroughs mainframe that ran the rudimentary calculations I had to make before I could work with the draftsman who toiled for months, drawing by hand the circuits I was designing. . . . That was the frustrating, cumbersome, counterproductive environment in which chips were designed."*

As the industry matured, semiconductor producers developed robust rules to guide the design of future generations of semiconductors. Designers determined what worked and what didn't work and developed guidelines for such things as the width and spacing of wires. They eventually embedded those rules in complicated electronic design automation (EDA) software programs. Electrical engineers began to spend less time learning detailed circuitry theory and more time learning logic and EDA software operations. Engineers who could manipulate EDA technology effectively became the new class of design experts.

The toughest problems in the high end of the industry continued to require expert knowledge and iterative and intuitive design. But less-skilled people could use machines with EDA software to perform tasks that previously required experts. Designers no longer needed to understand all of the circuitry behind the layout of chips. Advanced computer simulation made it possible for them to discard poor designs virtually and to modify other designs. What did this mean? Nonexperts could use the rules to design good-enough customized chips.†

*T. J. Rodgers, and Robert N. Noyce, "Debating George Gilder's *Microcosm*: T. J. Rodgers vs. Robert Noyce," *Harvard Business Review*, January–February 1990.

†Much of this is based on an interview with Randy Steck, formerly of Intel Corp., by the authors, Boston, Massachusetts, 26 February 2002. Steck taught us how experimentation uncovered specific rules that optimized the process design of logic (layout theory) and circuits (physical layout). For example, designers discovered forbidden zones, or patterns over the top of which wires cannot be run in a microchip for it to work. The development of rules like this one permitted designers to throw out prototypes that obviously violated some of the known rules for viable chip layout before they ever manufactured and tested them.

As discussed previously, efforts to wring every ounce of functionality out of the production process by having interdependent process steps and modular links between these steps means semiconductor production can take months. Historically, the product life cycles of Pentium processors were long enough, and customers' needs were predictable enough, that the speed of the production cycle just didn't matter.

But in a market with ever shorter product life cycles and custom-configured processors, fabs that take two months to deliver orders simply will not fly. The winners have to figure out how to deliver orders within days. Fabs need to move from a batch process with high WIP to a low-WIP, continuous flow process. The wafer fabs that succeed will be those that figure out how to make integrated circuits in a system similar to the one in which Toyota makes cars.

In this fast and configurable world, it will be the speed of the manufacturing process itself that is not good enough. Therefore, the manufacturing process must become interdependent and proprietary. Conservation of integration means optimizing the links between each step in the process. This will require *deoptimizing* each process step (and the equipment required for that process step) and making it configurable to optimize flow across those steps. Output from each stage needs to be as predictable and reliable as possible to avoid the need to spend time testing interim output. Achieving reliable output requires aligning process steps to obviate WIP and buffers. It means pulling back from the edge of what is possible. In all likelihood, the best means to achieve this is using known production technology (see "Implications for Next-Generation Technology"). Figure 7-2 shows how this interdependent process of the future (on the right) differs from the interdependent process steps (on the left) required in the world of the past.

In 2004, there were signs that this change was in fact occurring. TSMC and other foundries such as Taiwan-based United Manufacturing Company (UMC) had begun to move toward single-wafer processing (the semiconductor analog to the Toyota production system), meaning a single silicon wafer flows continuously through portions of the process without ever having to sit in large stacks of wafer WIP.

This change in the production process has obvious implications for companies that supply machinery to the fabs. The equipment that the fabs buy will have to become modular and configurable to support the fabs' ability to synchronize the flow of single wafers across the steps in

Implications for Next-Generation Technology

In 2004, building a leading-edge, efficient manufacturing facility required roughly $3 billion. These plants are known as 300 mm fabs because they produce chips on silicon wafers that are approximately 300 millimeters, or about 12 inches, in diameter. Because each wafer holds more chips, 300 mm plants promise greater throughput and lower unit costs.

Interestingly, the change in process architecture might mean that companies that place bets on next-generation technology could ironically fall behind. In other industries (such as automotive) where the conversion to low-WIP, connected-flow processes has been achieved, successful managers opted to employ the simplest, most reliable, and most proven equipment possible. In other words, equipment that used reliable, well-understood process technologies proved easier to configure into a synchronized process flow than did equipment that used cutting-edge technology, from which all of the kinks had yet to be worked out.

Furthermore, one of the big advantages of 300 mm plants is that they allow companies to increase throughput of high-volume products. However, a shift in customer demands toward short runs and customized products would obviate this advantage.

In this circumstance, we would expect the move to the latest design technologies that enable 300 mm production to happen in the most demanding tiers of the markets that value the higher performance and lower production costs. However, in other tiers of the market it is quite possible that this equipment will become dominant slowly, if at all.

the process. On the other side, foundries already are beginning to dictate the design rules to which chip designers must conform. Similarly, equipment providers will need to conform to standards established by foundries, whereas in the past it was the other way around.

How can companies such as Applied Materials and Tokyo Electron provide modular, flexible machines that are easy for foundries to custom configure to optimize flow? Over time, their machines will come to be comprised of modular subsystems. Subsystems will become interchangeable because well-defined standards will dictate how the parts must in-

FIGURE 7-2

Changes in an Overshot World

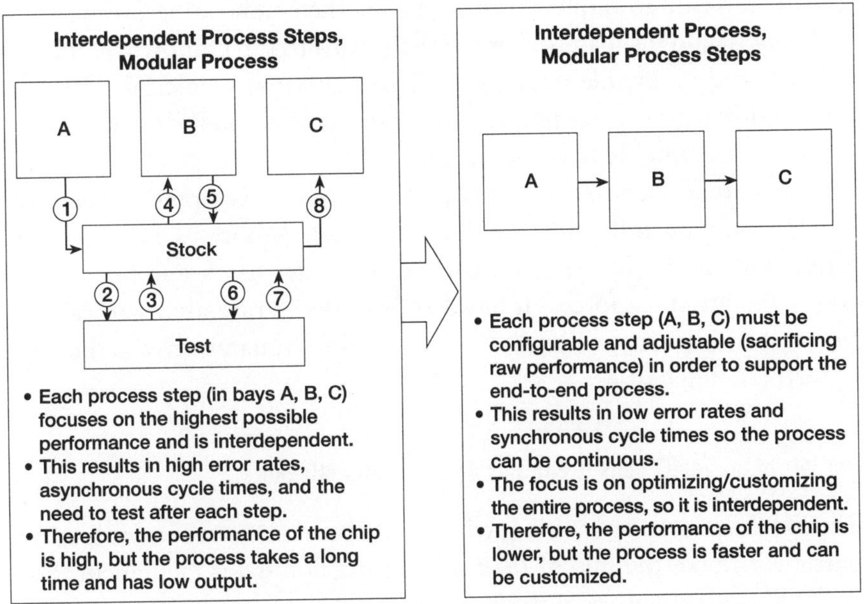

teract. The designs within these subsystems will become interdependent and proprietary. Companies that produce important subsystems therefore will see their fortunes improve, because what is not good enough will shift to components and subsystems.

Left unanswered is an important question: Would these sorts of developments be a good thing or a bad thing for Intel? In the next section, we look at looming competitive battles and show how our theories suggest this change *could* be a real threat. But we will assert that Intel might be one of the few operating companies in the world that has the capability to turn that threat into an opportunity.

Understanding Ultimate Victors: Evaluating Competitive Battles and Strategic Choices

If the industry evolves in a natural way, new entrants would have a great chance to drive industry change. Entrants that nail the new basis of competition in speed, customization, and convenience would become viable threats to leading IDMs. They would have to move fast to capture value. These entrants would grow based on asymmetric motivation; they would build unique skills to deliver their new value proposition; and they would ultimately sharpen their skills to the point where incumbents would be unable to respond. These entrants would have to continue following the right preparation regimen to avoid the temptation of creating a co-optable business model.

Why would incumbents not respond to these developments? Natural forces would push them to continue to pursue Moore's Law until their ultimate demise. However, we believe that Intel has a good chance of counteracting these forces. It has demonstrated a rare ability to capitalize on disruptive opportunities, which could ultimately give it the edge in forthcoming battles.

The Tale of the Tape Points to Entrants as Asymmetry Attackers

The first step in understanding who would be likely to win a competitive battle is to take the tale of the tape. As the dominant incumbent, how did Intel measure up as of the beginning of 2004?

Intel has strong resources. It has plants, scientists, patents, brands, and lots of cash to bring to battle. It also has powerful process-based skills. Remember, processes come from successfully and repeatedly addressing a task. Selling to large manufacturers, developing cutting-edge products, and managing a very complex manufacturing process are all examples of problems Intel repeatedly had to solve.

The next area to analyze is incumbent values that drive resource-allocation decisions. In chapter 2 we suggested starting with the income statement. Intel is very large and historically had high gross margins. From 2000 to 2002, Intel had average annual sales of close to $30 billion with gross margin percentages ranging from 49 percent to 62 percent. Another place to look is a company's past investment history. Intel historically invested a lot of money in next-generation production technologies. It placed a multibillion-dollar bet on next-generation 300-millimeter plants, presuming it would be able to gain scale economies and own the high end of the industry. A third proxy is a company's largest, most important customers. Historically, leading computer manufacturers were Intel's bread-and-butter customers.

Understanding Intel's skills and motivation shows why disruptive entrants such as Tensilica and Xilinx have a real chance to drive industry change. Intel almost *has* to invest in continuing to drive along the Moore's Law trajectory. Its processes have equipped it to solve the unsolvable and develop better products. Its most profitable customers are unsatisfied with current performance levels. Its values propel it to make the faster products that promise higher profit margins. Intel has almost no choice but to continue to improve. In doing so, however, it continues to create opportunities for upstarts. If this sounds like a classic innovator's dilemma, it should. The very thing that makes Intel great is the very thing that creates opportunities for the attacking firms.

Intel just doesn't have the natural motivation to play in the entrants' initial markets. Many of the markets for customized chips are just too small to prove interesting. To Tensilica, a $20 million market is enormous. To Intel, a $20 million market is a rounding error.

Entrants could settle on a business model that looks unpalatable to incumbents. For example, if entrants take advantage of established rules and rely on customers to do the design and foundries to do the manufacturing, they have no need for high research and development expenses. Indeed, as processors become modular in architecture and assembly of their constituted cores becomes rules-based, the design of processors will become a commoditized, marginally profitable stage in the value chain—just as the design of modular computers has become commoditized.

This commoditization of a formerly profitable stage of the value chain is why readers should pause before frantically calling their brokers to invest in fabless chip companies. If fabless firms are not careful, all of the value in the industry will skate to the customers who benefit from the convenience, the companies that provide the actual components for the customized chips, and the manufacturers. Modular assemblers typically find it difficult to capture value when what they do is undifferentiated and replicable. Fabless firms ought to make sure they closely control the point at which customization occurs so they can solve problems that allow them to generate value. Theory also suggests favoring companies who race to take low-cost business models up-market. Firms that are content staying in the lowest tier of the market will eventually see competition eat away at their profitability.

So the shield of asymmetric motivation looks secure. Could entrants create skills that run counter to incumbents' skills?

Intel's well-honed processes—which are almost unassailable competitive strengths in fights for undershot customers hungering for performance increases—might inhibit its ability to fight for customers clamoring for customized products. Its exacting manufacturing process could hamper its ability to deliver customized products. Its sales force could have difficulty adapting to a very different sales cycle. It would have to radically alter its marketing process. The VCE model predicts that operating "fast fabs" will be an attractively profitable point in the value chain in the future. The good news for IDMs such as IBM and Intel is that they own fabs. The bad news is that their fabs aren't fast. Entrants without legacy processes could quite conceivably develop better proprietary processes that can rapidly deliver custom processors.

Assuming entrants continue to improve, they eventually would begin pulling customers away from incumbents. The natural outcome would be for incumbents to choose flight, ceding new markets or the low end of the current market to disruptive attackers and seeking profits upmarket. Because the semiconductor industry still seems to have undershot customers, this strategy could be profitable and successful. But following this strategy would leave the fastest growing segments of the markets to new companies. And once the number of undershot customers dwindles, incumbents could be in trouble. By that point, attackers would likely have honed their unique skills. Incumbent response would come too late.

What Choices Could Change the Game

Two developments could dramatically change this game. The first is entrants getting greedy, forgetting their preparation regimens, and jumping too soon to the big markets. This would be obvious to observe and would lead to devastating consequences. Entrants bent on disruption must avoid fighting battles they cannot hope to win—trying to stretch underperforming products to reach large markets populated by very demanding customers. Analysts can continually check the entrants' management teams, investor needs, and strategy-making processes to assess how likely entrants are to stay on the right course.

The second development involves how incumbents respond to the growing threat of disruption. We mentioned that incumbents would be likely to choose flight. Could incumbents co-opt these potentially dis-

INDUSTRY SUMMARY

Semiconductors

Signals of Change: Decreasing premiums for next-generation products, emergence of specialists, and customers wasting functionality point to overshot customers.

Competitive Battles: Markets calling for customization are still relatively small compared to core markets. Delivering customized solutions requires unique skills.

Strategic Choices: Entrants must act fast to capture gains. Intel has shown the ability to recognize and react to disruptive threats.

---

ruptive business models? Theory suggests not. Co-option can be successful when an incumbent's processes facilitate it doing what needs to get done and its values prioritize co-option over other alternatives. The small size of new markets, the distinct business models required in those markets, and the unique skills necessary to serve those markets all point to a circumstance in which co-option would have a low probability of success. We would therefore expect incumbents who attempt co-option to end up cramming—forcing their one-size-fits-all product on markets calling for customization or trying to stretch underperforming customized products to demanding customers.

This assumes, of course, that incumbents such as Intel follow a natural course of response to the dilemmas they confront. If Intel's management can see these changes coming and take appropriate action, there is no reason why it cannot capture the growth that entrants would normally generate. Intel has shown a unique ability to wrestle with disruptive innovations; it is far down the path of earning its disruptive black belt. In response to AMD and Cyrix gaining market share with low-priced processors in entry-level computer systems, Intel created a separate organization in Israel to market its Celeron chip, blunting an attack from below. In 2003, it introduced its Centrino technology, which is optimized for the needs of the mobile worker. Centrino chips have less raw functionality but support wireless applications and battery life maximization.[17]

Its StrongArm logic circuit, deployed in its Hudson, Massachusetts, facility, carries the Intel banner in the disruptive custom-configured value network. If Intel continues to allow the technology to develop a business model and processes that are good at customization and rapidly shipping small unit volumes per order, its response could be successful. Remember, however, the VCE theory suggests money will shift away from microprocessors. Even a victory by Intel in this stage of value added might not mean much. The attractive money will be made by fast fabs. Right now, Intel fabs are configured to be slow. Financial analysts will pressure Intel to close its older fabs down and begin outsourcing production to firms such as UMC and TSMC. The better option, of course, would be for Intel to keep its fabs and make them fast—even offering fast delivery of chips designed by competing firms.

Using theory to analyze the future of the semiconductor industry requires looking beyond Moore's Law. If circumstances change, actions that were successful in the past might not be successful in the future. Companies mindlessly pursuing Moore's Law should not be mindlessly expected to triumph.

When leading firms lose the ability to receive price premiums for their next-generation products, overshooting has occurred. When smaller firms begin to get surprising traction in niche market applications where only the giants dared tread in the past, it signals the game is changing. Without the benefit of theory, these developments would likely be viewed as meaningless noise; with the benefit of theory, these developments can be viewed for what they are—harbingers of massive change.

A change in the basis of competition from functionality to speed-to-market, customization, and convenience will reverberate throughout the industry. The industry's value chain will have to organize around delivering increasingly quick turnaround times. Stages in the value chain that were proprietary and profitable will be commoditized. Stages that in the past were commoditized and marginally profitable will require proprietary products and services, resulting in much more attractive profitability.

The growth markets of the future will not be in today's monolithic one-size-fits-all product categories. Operating processes and cost structures that are assets today could be liabilities tomorrow. Fundamentally different business models will dominate the differ-

ent ends of the market. The big question is which companies will develop the necessary capabilities and organization structures to thrive in these markets.

The signals of change point to competitive battles in which incumbents could be on the wrong side of asymmetries. Unless they take aggressive action, today's heavyweights might be tomorrow's also-rans.

The semiconductor chapter has four general lessons:

- *The principle of conservation of integration helps describe how value chains need to reconfigure when the basis of competition shifts.*
- *Customers showing less interest in paying for performance increases and the entrance of nonintegrated firms that specialize at one stage of value added indicate overshooting.*
- *Overshooting enables companies to change the basis of competition, take advantage of rules-based design and manufacturing, and shift production closer to the customer.*
- *Disruptive business models are difficult for incumbents to respond to because they take unique advantage of motivation-based blind spots and skill gaps.*

Notes

1. Many of these ideas were first introduced in other publications. See Michael J. Bass and Clayton M. Christensen, "The Future of the Microprocessor Business," *IEEE Spectrum* 25, no. 8 (April 2002): 50–57; Matthew C. Verlinden, Steven M. King, and Clayton M. Christensen, "Seeing Beyond Moore's Law," *Semiconductor International* 34, no. 4 (July 2002): 34–39; and Steven Milunovich, "Technology Strategy; The Theory and Application of the Christensen Models," Merrill Lynch & Co., 27 March 2002. Our thanks to Matt Verlinden, Steven King, and Wilfred Pinfold for their input on this chapter.

The term *semiconductor* technically means any material that is neither completely conductive (such as copper) nor completely nonconductive (such as rubber). Silicon is the best-known semiconducting material, but at times other semiconducting materials such as germanium and gallium arsenide are used.

2. Moore based his so-called law on an observation he made as the director of the research and development lab at Fairchild Semiconductors in 1965. At that point, the number of transistors on a given area of silicon was doubling roughly every year, with no corresponding increase in cost. He predicted that this doubling would continue for the next ten years. In 1975, he predicted the pace would slow

so that processing power would double only every two years. He turned out to be wrong. The rate of improvement did slow after 1975, but only to eighteen months, quicker than Moore had imagined. Moore left Fairchild Semiconductors in 1968 to found Intel along with Andy Grove and Robert Noyce.

3. "The News of Radio: Two New Shows on CBS Will Replace 'Radio Theatre' During the Summer," *New York Times*, 1 July 1948.

4. For example, vacuum tubes burned out a lot. Therefore, appliance stores made a significant amount of money on postsales repairs. Transistor-based systems that were more reliable threatened these retailers' business models.

5. Francis Bello, "The Year of the Transistor," *Fortune*, March 1953, 132.

6. This story appeared in Clayton M. Christensen, *The Innovator's Dilemma: When New Technologies Cause Great Firms to Fail* (Boston: Harvard Business School Press, 1997), 232. Much of it was based on the personal recollections of Dr. Sheldon Weinig, retired vice chairman for manufacturing and technology at Sony.

7. Intel Corp., "How Transistors Work," <http://www.intel.com/education/teachtech/learning/transworks/> (accessed 26 August 2002); Semiconductor Industry Association, "Worldwide Semiconductor Shipments," <http://www.semichips.org/downloads/ACF8C.pdf> (accessed 21 August 2002).

8. Thomas Walter Smith, "Semiconductors," Standard & Poor's Industry Surveys, 18 July 2002; Intel Corporation, "What Is a Clean Room?" <http://www.intel.com/education/teachtech/learning/chips/cleanroom.htm> (accessed 22 August 2002).

9. Figures from Taiwan Semiconductor Manufacturing Company Web site, <http://www.tsmc.com> (accessed 25 January 2004).

10. Smith, "Semiconductors." Other important market segments include microcontrollers, dynamic random access memory (DRAM) chips, digital signal processing (DSP) chips, and flash memory. Microcontrollers are generally less computationally powerful than high-end microprocessors. They exert real-time control over other systems, such as automobile engines. High-density, low-cost DRAM chips provide basic functionality on computers, acting essentially as passive filing cabinets. DRAMs have many commodity-like characteristics—high volumes, high cyclicality, high degrees of competition, and limited ability for any firm to make supernormal profits for any period of time. DSP chips are stand-alone processors that power devices such as cellular telephones and DVD players. Texas Instruments is the leading DSP player. Flash memory has similar functionality to DRAMs, but items stay in memory even when power is turned off.

11. Intel is legendary for having a "copy exact" process. It might not know exactly why a specific technique works, or what variables must be carefully controlled for the process to have adequate yields. So once it has experimentally created a situation in which its process produces good output, it makes sure to have an exact replication throughout all of its manufacturing facilities.

12. It is important to note that this high level of WIP is not expensive because of raw materials alone. Highly skilled workers and costly equipment quickly add

value to those materials. And the high levels of WIP affect speed to market.

13. McKinsey Global Institute, "Productivity in the United States," McKinsey & Co., Washington, DC, 2001. Available from <http://www.mckinsey.com/knowledge/mgi/productivity/index.asp> (accessed 6 April 2004).

14. Common benchmark programs indicate these chips can perform more than a billion floating-point operations per second (one gigaflop) and, in some cases, more than two gigaflops.

15. Bass and Christensen, "The Future of the Microprocessor Business," 38. The 1997 National Technology Roadmap is accessible at <http://arch.cs.pku.edu.cn/users/chengxu/Org_web_ext/PDF_FILES/1997Roadmap_all.pdf> (accessed 5 April 2004).

16. Other leading programmable chip manufacturers include Altera, Actel, and QuickLogic. A logic chip processes specific operations or executes specific instructions to perform a variety of operations.

17. For more information about Intel's moves away from a pure focus on performance, see Don Clark, "Big Bet Behind Intel Comeback: In Chips, Speed Isn't Everything," *Wall Street Journal*, 18 November 2003.

CHAPTER EIGHT

HEALING THE 800-POUND GORILLA

The Future of Health Care

Do the theories of innovation apply to health care, or is it somehow different? Can health care providers overshoot customers? What do asymmetries look like? Who would ever want a "less good" product? How do nonmarket forces such as the third-party payment system affect disruption? What sorts of choices can firms make to maximize their chances of driving disruptive growth? What sorts of choices can the government make to encourage disruption?

In this chapter we turn to the 800-pound gorilla sitting quietly in the corner.[1] Health care is a vitally important industry that captures about one of every seven dollars spent in the United States. Is there any other industry where seeing the future is more important? Health care not only matters to the millions who invest in leading companies such as Johnson & Johnson, Merck, HCA, and Medtronic, it affects our lives and the lives of our children.

If you believed most newspaper stories in 2004, the health care industry was in crisis. The quality of care the U.S. health care system *could* deliver was beyond reproach. But often that quality got delivered late, or not at all. And costs were rising. Employers were pushing more of the cost burden onto their employees.[2] More people were uninsured, and a growing number of people expressed dissatisfaction. Harried doctors hustled patients through eight-minute appointments. Insurance plans told people what they could and could not do, and the "could nots" grew by the day. People complained that prescription drugs were too expensive. Books analyzing the industry carried such ominous titles as *Severed Trust* and *Bleeding the Patient.*

In the midst of the crisis, however, there is unprecedented opportunity for disruptive innovation to reshape health care. This chapter begins by providing three case studies that show how disruption in health care allows more people to consume more cost-effective health care in increasingly convenient settings. It shows how scientific advancements typically provide the fuel that enables these disruptions. Then we will turn to the future and look for signals of change that indicate innovators taking advantage of these forces and the strategic choices they can make to enhance their chances of success. Looking at health care also allows us to look more closely at nonmarket forces, to see what could change that would further encourage innovation.

Readers will note that this chapter does not analyze potential competitive battles in great depth. These battles are important. There will be winners and losers. But the real winner in the future we envision will be consumers, who will increasingly be able to manage their own health in convenient settings at lower costs.

The chapter ends by addressing a general criticism levied at theory-based analysis. "My industry is just different," some people say, "so your theories don't apply." Health care is undoubtedly complex. But good, circumstance-based theory is industry-agnostic. Even in an industry as unique as health care, the theories of innovation are a useful way to predict industry change.

Let us now show how the theories of innovation can help to identify the areas in health care that are ripest for new growth, to interpret market signals, to see unfolding changes, and to predict how innovation could reshape the health care industry.

Disruption in Health Care: Rules Allowing More Convenience and More Consumption

Disruptive innovations enable treatments that are less expensive and more consistently effective, allowing more people to consume more and higher-quality health care. This section discusses three such innovations:

1. Home pregnancy tests, which allowed women to see if they were pregnant in the comfort of their own homes
2. Blood glucose monitoring kits, which made it easier for patients with diabetes to treat themselves
3. Angioplasty, which allowed nonsurgeons to begin treating interventionally a condition that previously had been beyond their capabilities

Other examples of disruptive innovations in health care include free-standing ambulatory surgical and diagnostic testing centers. These disruptions share one common element: In each case, caregivers with less training became capable of providing effective care in more convenient, less expensive venues—care that historically had required expensive experts located in inconvenient, costly facilities.

Pregnancy Testing: From Rabbits to Kits

In the early 1900s, if a woman thought she was pregnant, what would she do?[3] She would go to her mother or to other friends who had been through several pregnancies. They would tell her the signs that indicated she was pregnant. She might go see a doctor, who would perform crude visual and tactile examinations to try to determine whether she was indeed expecting a child. Sometimes a woman could not be sure she was pregnant until she missed her menstrual period for several months and began to "show."

In the 1930s, what would she do? Amazingly, the answer involved a rabbit. The woman would go to a doctor's office. The doctor did not understand exactly how to identify a pregnant woman but knew that if the woman were pregnant, hormones in her urine would interact with an antigen in the rabbit's reproductive organs. To run this test, the doctor injected the rabbit with the woman's urine. The doctor would then kill the rabbit and examine its ovaries. Hence, the term "the rabbit died" evolved as a euphemism for pregnancy (even though the rabbit died regardless of the outcome!). Although the rabbit test was better than guesswork, it was expensive to perform because it required the use of special laboratory equipment, experts to read the results, and a large live animal colony.

What would she do in the 1960s? She would still go to an inconvenient central setting. But now doctors knew what to look for. A hormone known as human chorionic gonadotropin (hCG) rises at a certain rate in female blood and urine (known as serum) during pregnancy. Doctors could run a test that indicated how much hCG was in the women's serum. The first test was a radioimmunoassay (RIA) test.[4] The doctor would essentially combine a serum sample from a potentially pregnant woman, an antigen that was stuck to the bottom of a test tube, and a radioactive isotope that combines with hCG. Through a sequence of mixing and rinsing, the radioactive isotopes would bind to any hCG that had bound to any of the antigen. If the laboratory scientist could see any radioactive isotopes, she knew that hCG was present in the original sample.

This test was a great leap forward. It was accurate and could unambiguously determine that a woman was pregnant so she could roughly know her due date. It followed relatively straightforward rules. But it still required the involvement of expensive experts who had training in handling toxic radioactive substances. Over time, scientists improved the test and replaced radioactive material with enzymes, which through a series of reactions produced a visible color if any hCG was present in the original serum.

Fast forward to today. What does a woman do? She goes to her local drug store and buys a $10 home pregnancy testing kit. Home pregnancy testing kits first hit the markets in the 1970s. The first tests weren't very good: They looked like chemistry sets, were complicated to use, and were often wrong. Nevertheless, driven by strong consumer demand, the home pregnancy test's accuracy, sensitivity, and ease of use predictably and rapidly improved.[5]

The ultimate test came out in the 1990s: a one-step, no-mess kit so simple that it is almost impossible for users to make a mistake. Its accuracy approaches 100 percent. To use this test, a woman covers a test stick with a sample of her first morning urine. She waits five minutes. Voilà! If the stick turns blue, she's pregnant. If it remains the same color, then she is not. The woman still goes to the doctor to begin prenatal care, but the tests are so good and so accurate that they are just about never wrong. Americans now spend more than $200 million annually on home pregnancy test kits.[6] Unit consumption of these tests is many times greater than it was when testing was inconvenient and expensive. Scientific developments have driven this growth by allowing women to consume pregnancy testing in a setting that was unthinkable before.

Developments in Blood Glucose Monitoring

Developments in home blood glucose monitoring kits have similarly allowed patients with diabetes to accurately monitor their blood sugar levels without a physician's assistance.[7] Before 1980, diabetics could either use an inaccurate urine test or go to a centralized location to use expensive laboratory testing machines manned by technicians. The home blood glucose monitoring tests changed this. Users now prick their finger and squeeze a drop of blood on a test strip that is translated into a numerical reading by an electronic monitor.

Today, patients can bring these miniature monitors wherever they go, allowing them to manage for themselves dimensions of a disease that historically required significant professional involvement. They can now run tests several times a day to precisely monitor their blood sugar levels and take corrective actions when necessary. These developments drove companies that made the expensive testers from the marketplace. Patients who use these convenient glucose monitors are often more knowledgeable about managing diabetes than their physicians. In fact, they rarely need to see their physicians at all. Endocrinologists, as a result, face lower demands for their services. But patients with diabetes are far better off.

Healing a Heart

Finally, consider what happens when you have chest pain, known as angina. Angina is sometimes a symptom of coronary artery disease, which occurs when atherosclerosis results in plaque that blocks (or occludes) one of the arteries that carry blood to the heart. If you had an occluded artery in the 1950s, it led to tiredness and, often, tragic death. The best prescription was bed rest. Your only hope was to minimize the stress on your heart and, consequently, the amount of oxygen the cells would need to receive through the narrowed vessel.

Doctors then began experimenting with a different approach—taking a vein from the leg and inserting (or grafting) it so that it allowed blood to "bypass" the occluded artery and maintain flow within the heart. Coronary artery bypass graft surgery (CABG, today known as bypass surgery) is one of our time's medical miracles. It is also risky and expensive, involves weeks of recovery time, and requires highly skilled cardiac surgeons using complicated equipment. Because the operation is so traumatic, doctors reserve it for the direst cases, in which surgery makes the difference between life and death.

Another option emerged in the 1970s when Andreas Gruentzig successfully introduced a form of treatment known as angioplasty. Angioplasty is far less invasive and significantly easier to perform. Instead of having to split open the chest, arrest the heart, and graft an artery, cardiologists (who were not trained to perform this invasive surgery) puncture a small hole near a patient's groin. They carefully insert a catheter and maneuver it through the blood vessels to the heart. Once the catheter is correctly positioned in the occluded artery, the cardiologist inflates a balloon attached to the end of the catheter. The balloon dilates the artery by forcing the atherosclerotic plaque back against the walls of the artery.

Angioplasty is a disruptive innovation—much simpler and cheaper to perform than bypass surgery. It does not require the specialized training of a surgeon to execute. Because it is less invasive, it is less risky and requires much less recovery time. It allows patients with partially occluded arteries, for whom bypass surgery had not been an option, to begin receiving interventional therapy. And cardiologists who previously had to refer heart patients to cardiac surgeons can now do treatment themselves.

Like all new-market disruptions, the first angioplasty treatments were not as efficacious as CABGs and could not be used on the most severely occluded arteries. They suffered from a common problem known as restenosis, in which an overproduction of cells in the months following the procedure resulted in a repeated occlusion of the artery at the site of the original plaque. But relative to no interventional treatment at all—the only alternative available to patients with partially occluded arteries and their cardiologists—it was *much* better.

Angioplasty has predictably improved since Gruentzig's first successful demonstration. The adoption of stents—cylindrical mesh cages that remain after the doctor removes the catheter—in the late 1980s considerably lowered the rate of restenosis. Recent inventions such as stents coated with a slow-releasing drug that prevents postprocedure blockage will further lower restenosis.

Angioplasty procedures grew at a compound rate of almost 40 percent between 1979 and 1999. The number of angioplasties exceeded the number of CABGs in 1997. CABGs declined between 1997 and 2001.[8] Heart surgeons are beginning to feel the pain as cardiologists disrupt them. But, as with all disruptions, the result is that more people are much better off.

Disruption Makes Treatment Cheaper and More Convenient

In each of these three cases, disruptive innovations powered two changes. First, providers shifted closer to consumers. Health care providers form a pyramid. At the bottom, of course, are patients. Next are nurses. Above nurses are generalists. Then there are specialists. At the top are subspecialists. Each tier up the pyramid is smaller, more highly trained, and more highly paid.[9] A *provider-level* disruptive innovation, displayed on the left side of figure 8-1, moves the ability to provide competent care down the pyramid, allowing patients to move into the realm of nurses, nurses to move into the realm of general practitioners, and general practitioners to move into the realm of specialists.

Developments in the treatment of heart disease illustrate this pattern. Angioplasty allowed cardiologists over time to begin to disrupt cardiac surgeons.

Provider-level disruptions begin like traditional new-market disruptive innovations. They expand treatment by allowing less-skilled providers to reach nonconsumers. The provider begins treating conditions that are not quite severe enough to justify treatment by the higher-skilled provider.

Provider-level disruptions tend to make people nervous. They say, "Wait a minute. What about the specialists, what happens to them?" The disruptive innovations actually free specialists to work on solving even more challenging medical problems. In other words, disruption does not mean that our best, most highly trained physicians will have nothing to do. They can focus on the toughest problems that best match their skills and training.[10]

FIGURE 8-1

Disruption in Health Care

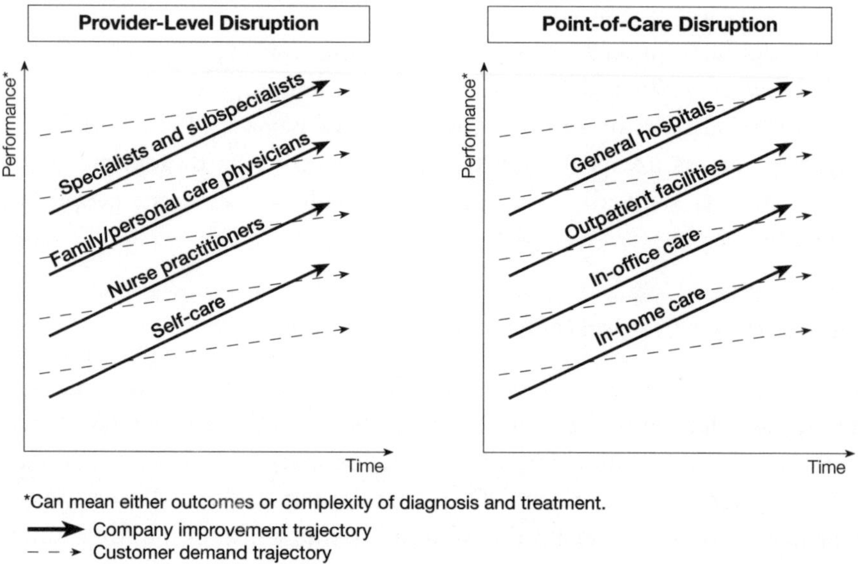

The second part of disruptive change is that it moves treatment to more convenient, less costly settings. This is a *point-of-care* disruption (illustrated on the right side of figure 8-1). Teams of specialists in general hospitals treat the most demanding cases. Consumers can treat the least demanding cases at home. Recall how successive innovations in pregnancy testing moved initial tests from the hospital to our neighborhood doctor to our own homes. This is one manifestation of a general pattern. Technological progress enables us to receive care in the doctor's office that previously we needed to receive in a hospital, and it eventually enables us to receive care at home that previously was only available in the doctor's office. Disruptive innovations enable caregivers and venues of care that are less expensive to move up-market, doing progressively more sophisticated things.

What typically enables the progression down both chains toward self-care by individuals? The development of rules to guide treatment. As hard problems become rules-based and routine, progressively less-skilled people can do in convenient settings what previously required expert skill in expensive, complicated hospitals.

Great Science: The Fuel of Disruption in Health Care

Initially, when things aren't well understood and diagnosis and treatment are more art than science, experts need to treat conditions in expensive hospitals, where they must diagnose and treat in an iterative, problem-solving mode. Through their training and experience, these experts have the ability to recognize patterns and have an intuition for what solutions might work. Then, little by little, disease by disease, scientific advancements begin transforming the provision of care from an art to a science. This transformation gradually allows caregivers with less training to follow clear rules to solve problems that previously had required the judgment and skill of the most highly trained experts.

When it is not clear what causes a disease, or when caregivers can only characterize a disease in broad, symptomatic terms such as depression, asthma, or leukemia, competent care requires the judgment of certified specialists. But when diseases can be unambiguously diagnosed, standardized best-practice treatments can be defined. This is what enables less expensive caregivers with less training to move up-market, providing better care at lower costs than previously was possible.

A simple two-by-two matrix, as shown in figure 8-2, displays this phenomenon. The horizontal axis depicts the clarity of the rules that govern a solution, representing a continuum between vague or nonexistent rules (on the left) that require unstructured problem solving and crystal-clear rules (on the right) that permit rules-based solutions. The vertical axis depicts the skill required to implement a solution. Solutions that require a high degree of training lie on the top of the vertical axis; solutions that require little training lie at the bottom.

At the far left of the horizontal continuum are medical disorders that lack diagnostic or therapeutic protocols that predictably work. These conditions include neurological problems that are hard to diagnose because the brain is so complex and so delicate, conditions that are complicated by comorbidities (a term that means a patient has multiple interdependent diseases that interact unpredictably), and "wastebasket" diagnoses (such as chronic fatigue syndrome) that are given only after a series of tests eliminate any other possibility.

At the far right are medical disorders for which clear rules direct diagnostic and therapeutic protocols. Accurate data yields an unambiguous diagnosis, indicating a proven therapeutic strategy. Strep throat, for

FIGURE 8-2

Migration in Disease Treatment

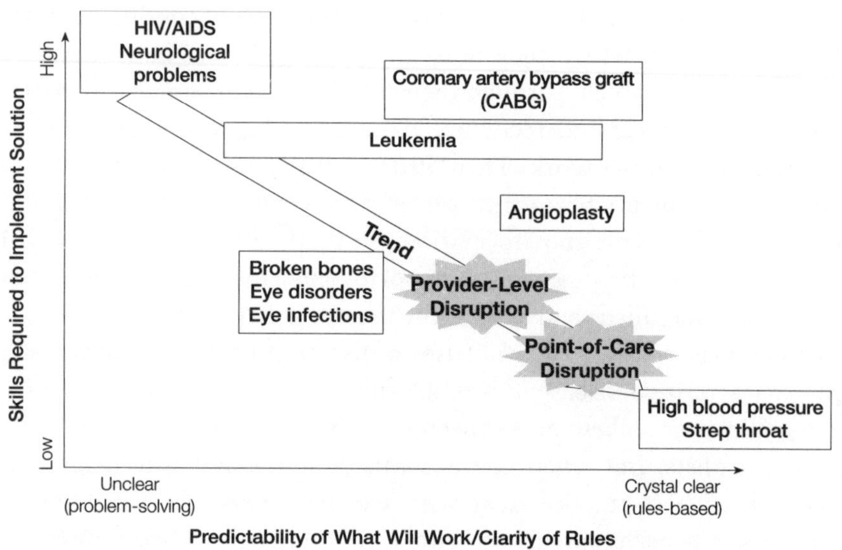

example, has recently shifted into this cluster. A simple test strip can yield an unambiguous diagnosis of strep throat.

Problems typically start in the upper left-hand portion of figure 8-2. Moves from the upper left of figure 8-2 to the bottom right enable disruption.

One important note. Even though implementing a solution might be extremely rules-based, it still might require expert skill, placing it in the upper right portion of figure 8-2. For example, CABG surgery is largely standardized. The rules a surgeon needs to follow are clear. But the procedure is still quite complicated to perform. For most cases, however, the development of clearer rules (a move to the right along the horizontal axis) catalyzes the development of simpler procedures that enable the up-market shift of a less-skilled tier of providers (a move down the vertical axis).

With an understanding of what disruption in health care is and what drives it, we turn to looking at the types of innovations that could bring disruptive change, the choices firms can make to realize this potential, and the impact of nonmarket forces.

How to Spot Signals of Change and Evaluate Strategic Choices in Health Care

As scientific developments continue, the possibility exists for innumerable new-market disruptive innovations that enable nonproviders to treat nonconsumers and for low-end disruptive innovations to create low-cost ways to deliver care.

Identifying nonconsumers is easy. People who have illnesses that go untreated are nonconsumers. People who must wait for a condition to get severe enough to treat are nonconsumers. People who receive treatment in hospitals do not consume care outside of hospitals. And so on. There is a clear parallel concept here. If an outpatient clinic can't provide care that a hospital can provide, it is a nonprovider. If a nurse can't do what a doctor can, she is a nonprovider for that specific condition.

When a physician with twelve years of scientific and medical training diagnoses your strep throat with a simple yes-no test strip, you are only using a tiny fraction of the physician's capabilities that you are paying for. When you receive a simple, low-risk procedure in an academic medical center that is equipped and staffed to treat the world's most complicated disorders, you are using only a fraction of the capabilities that you are paying for. In either case, the level and cost of care that you receive overshoots what you are able to use.

One reason why so little progress has been made in improving the cost and effectiveness of our health care system is that people use the term quality indiscriminately. In the upper-left regions of the disruption diagram (corresponding to the upper-left region of figure 8-2), when the efficacy and reliability of diagnosis and treatment are not yet assured, quality health care means getting the most effective, reliable care possible. But in the lower-right portions of the diagram (corresponding to the lower-right region of figure 8-2), when efficacy and reliability are more than adequate, patients define quality very differently. In this circumstance, quality means getting the required care as quickly, conveniently, and inexpensively as possible.

To help show what sorts of developments would be signals of change, we will discuss two potential disruptions in progress as of early 2004:

1. Drug treatment of heart disease, which is a new-market, provider-level disruption

2. Nurse practitioner–based primary care, which is a hybrid new-market and low-end point-of-care level disruption

New Market: Healing a Heart Continued

There is no better example of a massive new-market opportunity than the genomics revolution. What genomics, defined as the study of genetic maps of whole organisms, really seeks to do is to comprehensively classify all of the genetic circumstances in which humans might find themselves. Medical treatment is, in fact, a theory. If a doctor can classify precisely what condition you have and understand what that means, then the doctor can prescribe something that treats it. The problem is that we often don't have a good categorization scheme. We lump together multiple forms of cancer because there is no reliable way to distinguish one form from another. The problems are too interdependent.

Over time, experimentation tends to address this deficiency. People used to think leukemia was a single disease. Because different people reacted differently to the same therapy, we relied on the experience, intuition, and problem-solving skills of our very best oncologists, or cancer specialists. With our improved understanding of the genome, we have come to realize that what we called leukemia was really at least six separate diseases, each identifiable by a specific genetic pattern. The reason any given treatment regimen would work for some patients but not for others was that they had different diseases, which can now be diagnosed precisely by examining their genetic patterns. In the near future, precise diagnosis will allow providers to develop and follow a standardized, optimal treatment regimen for each specific disease with confidence that it will work.

Returning to our discussion of heart treatment, scientists are working furiously to determine the genetic makeup common to heart disease sufferers. If they could crack the genetic code, people at high risk of heart disease would start taking the appropriate drug at an early age. This move from treatment to prevention would ultimately allow each of us to disrupt specialist providers. Similar advancements in home monitoring and treatment devices would create further disruptive growth by allowing us to proactively manage our own health and prevent future ailments.

This is nonconsumption in health care: reaching people who previously had no way to consume treatment, moving from treatment to prevention, and bringing treatment to a new context.

Low-End/New-Market Hybrid: Nurse-Based Care

Many current means of treatment just plain overshoot customers' needs. The move of conditions such as ear infections, strep throat, and sinus infections to a pure rules-based mode means that going to a doctor overshoots our needs.[11] Imagine a mother who wakes up to find that her young son has an earache. She knows all of the symptoms. She knows the prescription drug she needs to get. But she cannot write the prescription herself. She thinks to herself, "Great. I know what I need but I have to take three hours out of my busy day to sit in my doctor's office while he tries to work me in." What is not good enough in this situation is not necessarily the skill of the practitioner. It is the convenience of the delivery.

If the mother lived in Minnesota, she could go to a kiosk located in her local Cub Foods supermarket or Target store. A company called MinuteClinic runs the kiosk in which a friendly nurse practitioner or physician's assistant can diagnose and treat a set menu of disorders. No doctor is necessary, and no appointments are required. The vast majority of patients are in and out within fifteen minutes. Patients can pick up prescriptions at the in-store pharmacy. The technologies these nurse practitioners use are not unique. But when wrapped in a low-cost, convenient business model, these technologies become an essential part of a disruptive delivery mechanism that improves the quality of care for common problems, in ways that patients define quality.

Just about any condition that highly trained experts treat in an inconvenient setting is a candidate for disruption. Any innovation that brings treatment closer to the individual consumer, if implemented correctly, has the potential for tremendous growth. This growth has elements of low-end disruption because it targets overshot customers and elements of new-market disruption because it enables nonproviders to treat conditions in a new context.

People in the health care system often get nervous about nurse-driven business models. They say, "What if the nurse misses *X*, *Y*, or *Z*, or she can't treat *X*, *Y*, or *Z*?" The point is not that nurse practitioners today should treat *all* conditions. In *certain* circumstances where there are well-developed rules, providers with nurse practitioner training are more than adequate. Studies have shown that nurse practitioners or physician's assistants can treat certain conditions as well as—if not better than—doctors.[12] In fact, patients appreciate the personal care nurses give. Nurses have the right sort of training and skills to meet customer demands on the dimensions of quality that are not good enough (such as convenience).

Nurse-based care requires that nurses refer patients to the next level in the delivery chain when they encounter problems that they cannot solve following well-developed rules. This is no different from the expectation we place on primary care physicians. We trust that the primary care physician will know what she does not know and refer us to a specialist when appropriate.

Disruptions That Reshape Industries Make Lives Better

Innovations such as drugs to treat heart disease or nurse-based primary care that reach nonconsumers or provide new benefits to overshot customers could reshape the health care industry. What's more, these sorts of innovations meet the jobs people are looking to get done in health care. The health care system is great at doing the "I am in desperate need of a difficult fix" job. But many of us are looking to get other jobs done.[13] For example, so-called disease management companies that aim to help treat patients with chronic diseases have struggled to gain traction over the past two decades. One reason is that most have framed their challenge as treating an entire disease. Instead of understanding the jobs customers are seeking to get done, they have developed detailed scientific expertise so they can handle the most complicated cases. The real job that many chronically but not seriously ill patients are trying to get done is to not have to interact with the health care system at all. Job-based segmentation, as much as disruption, is a key to successful growth creation in health care. Much of what companies spend creating and running disease management companies is wasted, because it improves along the trajectories that are not what customers want.

What are the real jobs many of us are trying to get done? Things such as "keep me well," "let me manage my own health," and "give me the prescription I know I need as quickly as possible." Innovations that make it easier for consumers to do these sorts of things will find the most success. Interestingly, the best delivery mechanisms to fulfill at least some of these jobs might bear little resemblance to what we traditionally think of when we think of health care. A nutritionist at our gym or a massage therapist might allow us to fulfill these jobs better than a doctor ever could.

Disease management companies that choose to focus on these jobs could enhance their disruptive potential. The next section points to other strategic choices companies can make to increase their ability to harness disruptive developments.

What to Look for in Disrupting Firms: Letting Nonproviders Reach Overshot Customers and Nonconsumers

A firm's most important decision—and the one it gets wrong most frequently—is the selection of its initial target market.[14] Firms seeking disruption in health care have the best chance of success when they enable nonproviders to reach overshot customers or undemanding nonconsumers.

Firms really have to make two choices when choosing a target market: choosing a target condition and choosing a target caregiver. First, firms have to decide which conditions and patients they ought to target. There is always great temptation to target the sickest, hardest cases. These patients are easy to identify and willing to pay a lot to ease their pain. Improving the lives of people suffering from debilitating diseases is certainly a good thing. But it will not transform the health care industry into one characterized by affordability, convenience, and effectiveness. These things come through disruption.

Most medical companies focus their investment on developing sophisticated technology that helps the most highly skilled professionals provide even better service. Stretching to solve unsolvable problems involves complicated, expensive, untested technologies. This feeds into concerns that cost, quality, and access are mutually exclusive goals. The theory of disruptive innovation suggests augmenting those investments with investments targeted at developing sophisticated technology whose purpose is to simplify delivery and treatment and render them foolproof.

If the industry focused more energy on coming up with disruptive ways to treat the more mundane things that affect our everyday lives—and then improving upon these treatments—more people would get higher-quality, more convenient treatment.

For example, consider the choices made by Sonosite, a spin-off of ultrasound giant ATL Ultrasound (which has since been absorbed by Phillips). Large players such as GE, Toshiba, and Phillips have dominated the ultrasound industry, providing high-powered machines that let doctors use reflected sound waves to produce pictures of organs and other structures inside the body. These machines sell for several hundred thousand dollars and require skilled technicians. Instead of trying to compete head-on against these giants, Sonosite has sought to find ways to bring ultrasound to places where it was historically impossible. In 2002, it introduced a three-pound, hand-carried ultrasound monitor (the leading players' monitors are all stationary). This monitor enables new applications. For instance, physicians in emergency rooms can use the device to take "quick looks." Doctors and nurses can use it to guide catheter insertion—a procedure that historically relied on guesswork and had relatively high error rates. The device's quality was vastly inferior compared to the $300,000 machines sold by industry heavyweights. But it allowed a new class of physicians to use ultrasound technology in new ways.[15]

After picking a target market and developing an innovative way to treat that market, firms next have to pick a target caregiver. Of course, if a firm develops a product for an individual consumer, the decision is easy! But companies often try to force simple, convenient means of treatment on highly skilled providers. These efforts often fail because the innovation pales in comparison to the provider's well-established, high-performing way of doing things. Firms have the best chance of success when they sell to nonproviders whose lack of specific skills or training locks them out of a market. The nonprovider's frame of reference is nothing. They will embrace the product and begin the classic march upmarket. Notice how Sonosite, for example, did not target the most demanding ultrasound users such as cardiologists or obstetricians and gynecologists.

This approach harnesses asymmetries. The lower-tier provider typically begins by treating patients not sick enough to see the higher-tier provider or by performing procedures that the higher-tier provider views as mundane. The lower-tier provider is motivated to add what to him or her appear to be more high-fee procedures. The higher-tier provider is motivated to drop what to him or her appear to be the low-fee procedures.

By making such choices, these providers can build asymmetric skills. As they treat lower-level cases, they develop unique skills in delivering care using a new product or procedure. When they can perform well enough to begin making inroads in established markets, incumbents do not have the skills to compete. Established providers would need to build what the attackers had already built. They would have to jump down to the beginning of the learning curve to develop the necessary skills.[16]

Two other choices can help firms nurture disruptive innovations. First, firms basing their businesses on different care-delivery models ought to minimize their integration with existing value networks. Using traditional insurance mechanisms, for example, could mean using a business model that makes sense to established providers while conforming to insurance company role limitations that dictate what each class of caregiver can do. A tight connection might look good in the short term but might ultimately limit the degrees of freedom to develop a legitimately different value proposition.

Second, firms must carefully consider their hiring decisions. In the industry, one school of experience that many look for is a high level of medical training. But that training is not particularly helpful for a company looking for senior managers to oversee a low-end disruptive business model based around process excellence. Having a medical background helps, but having experience in an industry that has to get a process right over and over again is even more critical.[17]

Companies that make these sorts of decisions have the greatest chance of finding the footholds discussed at the beginning of this section. Of course, nothing is ever quite so simple. The government has to approve new drugs. Nurse-based care companies have to fight the fact that the payment mechanisms make a low-cost business model look expensive. Firms making all the right choices might still be locked out of a market. Nonmarket forces are everywhere. The next section shows how to use the motivation/ability framework introduced in chapter 1 to help understand these forces.

Nonmarket Forces Play a Critical Role in Industry Change

Not surprisingly, nonmarket forces play a very important role in shaping innovation in health care. Because the health care industry is so vast and complicated, understanding how these forces work is a daunting task. One of the goals of our ongoing research is to understand exactly how to design a more disruption-friendly system. Nonetheless, at a high level, the concept of motivation/ability can help explain developments that will influence innovation.

In this section, we will walk through figure 8-3. We will first show examples of parts of the health care system that are in the hotbed; then we will explain how things such as the third-party payment mechanism dampen motivation; and finally, we will illustrate how factors such as

FIGURE 8-3

Nonmarket Forces in Health Care

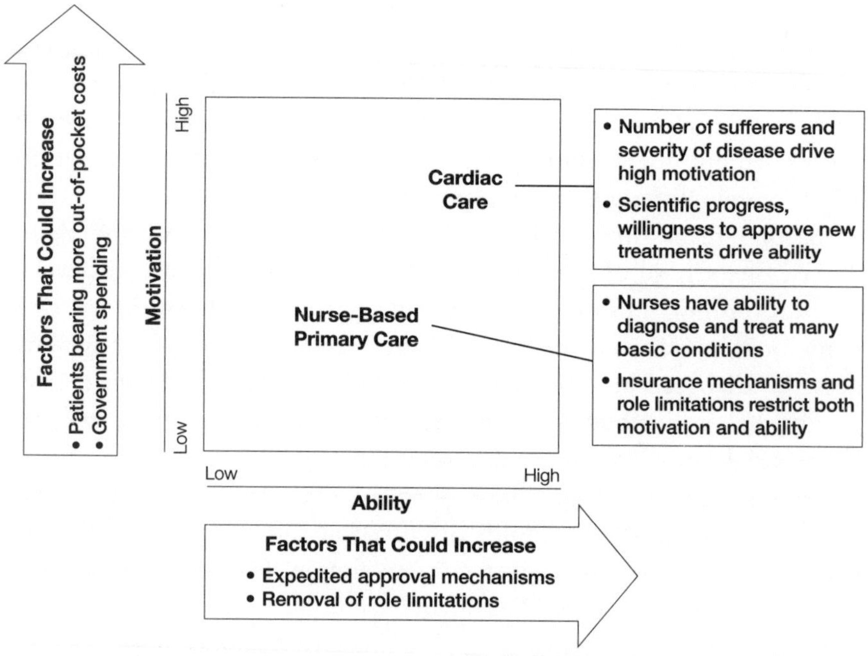

licensing rules constrain ability. Our most important message is that, historically, our system has shielded consumers from making the trade-offs that are central to disruption. Actions that expose consumers to trade-offs could raise motivation and result in a rash of disruptive innovations.

Remember, the motivation/ability framework holds that innovation flourishes when motivation and ability exist in abundance. For instance, scientific progress has moved cardiac care into the upper-right quadrant of figure 8-3. Better understanding of human physiology and the development of techniques such as angioplasty grant cardiologists the ability to solve previously unsolvable problems. Demand is high. The government is willing to approve devices and drugs, and insurance companies will reimburse many forms of treatment. The treatment of heart disease is teeming with innovation; firms are racing to develop new devices to improve angioplasty and novel pharmaceutical treatments to improve the condition of the millions of Americans who either have or are at risk for heart disease.

But not all forms of disease or all innovations are in the upper-right hand quadrant. At a basic level, the third-party payment mechanism dampens the motivation for some classes of innovation. Regulatory approval mechanisms and licensing rules that define what each industry participant can treat constrain ability.

The third-party payment mechanism dampens the motivation to develop disruptive business models by shielding companies and individuals from trade-offs. Trade-offs are important because consumers typically only choose to consume disruptive innovations when they pay directly for goods and services. Without a cost or convenience trade-off, savvy consumers will often choose the most costly option. Let's say your child has an earache. For the same copayment of five dollars, you can choose to have either a doctor or a nurse practitioner examine her. If both options were equally convenient, most people would choose the doctor, which is actually the more expensive option. But what if the doctor says, "Treatment *A* works 95 percent of the time. You can administer it yourself and it costs $10. Treatment *B* works 98 percent of the time. It is complicated to administer and it costs $1,000." Faced with a trade-off, patients might prefer the less costly disruptive alternative.

Third-party reimbursement systems sap motivation for innovation—particularly disruptive innovation—out of the system. When cost is not a consideration, patients always choose the higher-performing technology. When consumers are not encouraged to trade off some features that are marginally useful to them in exchange for lower prices (which they have no incentive to value) and greater convenience, they might not even realize that an existing technology overshoots their needs. In other words, industries in which consumers do not face a trade-off between performance and cost are undisruptable. The payment system pushes innovations such as nurse practitioner–delivered primary care toward the bottom half of the motivation/ability framework.

Similarly, nurse practitioners have the capability to treat many simple conditions, but insurance systems typically refuse to reimburse nurse-delivered care. This decreases the motivation to create a business model based on nurse practitioners delivering primary care. Without insurance reimbursement, consumers would have to bear the expenses themselves.

What could change? Companies have begun to explore offering their employees defined contribution health care plans. In these plans, companies give employees a fixed sum of money to manage their own health care. Interestingly, consumers might ultimately benefit from having to pay more out of pocket. It could increase the motivation of entrepreneurs to develop simple, low-cost disruptive innovations. The result could be more people consuming more health care at lower prices. This is an extremely important point. Although the commercial and emotional need to cure cancer and other fatal diseases is huge, we believe investments in technologies that enable individuals, nurse practitioners, and physician's assistants to provide more care would lead to greater industry-wide impact and greater growth opportunities for firms. Yet we observe that the Food and Drug Administration and the National Institutes of Health do not focus resources on the exact things that could transform the health care system. We predict that removing the third-party payment mechanism will take the cap off the motivation to innovative disruptively, increasing the rate with which the health system is transformed.

Firms could have all the motivation in the world but not be able to innovate because of barriers to ability. Two specific nonmarket developments could increase ability and lead to more innovation:

- Relaxing *role limitations* that let nurse practitioners treat more conditions could help innovators develop more disruptive primary care business models.
- Relaxing *approval mechanisms* could help innovators develop devices and drugs that do not have to stretch to meet the needs of the most demanding customers. The government put many regulations in place when conditions were not well understood. The regulations were well grounded and served the public interest well. They remain appropriate for conditions that are still in the problem-solving portion of figure 8-2. However, they are less appropriate for conditions that are in the rules-based portion of figure 8-2. Here, well-intentioned regulations can limit access to highly convenient, low-priced solutions.

Developments in the other direction—tighter role limitations or stricter approval processes—would slow the pace of innovation.

INDUSTRY SUMMARY

Health Care

Signals of Change: Potential signals include: any move closer to the consumer, actions to increase motivation (such as consumers paying more), or developments that increase ability (such as scientific breakthroughs or relaxed role limitations).

Competitive Battles: Procedures that look mundane to highly trained providers often look exciting to less highly trained providers.

Strategic Choices: Disruptive entrants need to avoid temptation to treat the sickest, to target highly skilled providers, or to overvalue medical expertise.

Watching these nonmarket forces provides deeper insight into how change is likely to unfold. Before moving on to the next chapter, we need to address one overarching issue. This chapter's final section addresses a common reservation many people express about theory-based analysis—that the industry in question is "just different," limiting the usefulness of broad theory.

Health Care: Less Different Than You Might Think

The forces of innovation affect every industry. Every industry, of course, has its unique nature, which affects the way in which innovation unfolds. Health care's value network affects where innovations are most likely to first take root. The problems industry players must repeatedly solve determine their collective skills. Firms' business models affect what they will find disruptive. We must take all of this into account when using the theories of innovation to gain insight into the future. But like every other industry, health care contains nonconsumers, undershot customers, overshot customers, asymmetries, freestanding value networks, and all of the circumstances that present opportunities for innovation to drive industry change. As long as we can identify the right signals, the theories of innovation remain a useful way to see the future.

People press on. "Okay, I'll grant you these circumstances exist," they say. "But people *shouldn't* want disruption in health care. In this context, it is a bad thing. After all, we are not talking about computers or telephones. If you get it wrong, people die." Their arguments tend to fall into three categories: doctor knows best, risk-averse patients, and social good. Let's look at each argument.

Doctor Knows Best or Doctor Knew Best

People often point to the overwhelming need to have a delivery system based around highly skilled, highly trained doctors. A doctor-centric system, they argue, requires that innovators focus on sustaining the current delivery mechanisms. If this argument were true, the existence of nonconsumers or overshot customers would not really be relevant.

Doctors historically did need to be at the center of the system for two reasons. First, there is *uncertainty of need*: People just don't know exactly when we need to consume health care. Second, there is *information asymmetry*: Medical providers just know much more than we do. But what was the root cause or driver of that paradigm, and has that root cause changed?

As medicine grew from a craft to a science over the past fifty years, patients had little choice but to exercise blind faith in their doctors' opinions. Doctors had specialized medical knowledge to which patients were not privy. We needed a trained professional to guide our decisions about when and how to consume health care.

But little by little, the availability of information over the Internet and affordable self-diagnostic tests are allowing individuals to resolve the uncertainty of need issue. And for better or worse, today the doctor doesn't always know best. Doctors can rely on outdated information. A patient who is willing to invest the time can often know as much or more than their doctors can about a specific condition. Thanks to the information revolution, patients have access to a trove of medical knowledge at the click of a mouse. Health-related information is one of the most sought-after types of information on the Web.[18] Some hold that patients want the doctor to tell them what to do. For some people raised in the days of asymmetric knowledge, this is true. However, many patients we talk to find the availability of information and self-diagnostic tests to be empowering. Studies of patients' attitudes toward health reveal that fewer than 20 percent of people think doctors are in charge of the health care relationship.[19] These kinds of changes are why disruption is more conceivable than ever before in health care.

The Myth of the Risk-Averse Consumer

Next, some suggest that patients will reject any untested, relatively simple innovation because they are unwilling to take risks with their health. This would mean that patients would be unwilling to risk their health by receiving care rendered by less-skilled people in lower-cost settings. But how does this purported risk aversion fit with the obvious risks that many of us take every day? Many of us still forget to fasten our seatbelts when driving. Far too many of us indulge in behaviors scientifically proven to be unhealthy, such as smoking, consuming excessive amounts of alcohol, and eating too much fatty food. People actually take *lots* of risks with their health. They climb on rickety ladders. Even extreme (and extremely risky) sports such as hang gliding and bungee jumping have become increasingly popular.

Furthermore, consumers spent close to $100 billion on alternative medicine in 2003. The willingness of a significant proportion of Americans to pay with their discretionary dollars for homeopathy, naturopathy, and relaxation therapies without consulting their doctors indicates that a growing number of patients were willing to place their trust in remedies that have not been approved by their doctors. Every day, in many ways, people demonstrate a willingness to take risks with their health, providing another piece of support for why disruption is a feasible strategy.

If Health Care Is a Social Good, Then We Should Encourage Disruption

People argue that because health care is a social good, everyone deserves the best. Disruptive innovations typically start out with less raw functionality. Since they are not the best, the argument goes, they should be discouraged.

The key to this issue is the definition of quality in health care. From society's viewpoint, costly and inaccessible health care is not high quality. Any argument based on social good should be *for* rather than *against* disruption. Historically, disruptive innovations increase social welfare by providing people with a highly desired bundle of attributes. Remember, disruptive innovations offer relatively low performance along *some* metrics, but they introduce better performance on others, such as simplicity and convenience. Furthermore, by lowering costs, disruptive innovations would help achieve any social goals by increasing access to high-quality health care.

The health care industry has its share of problems. It seems trapped in an unbreakable paradox: Treatment grows ever more expensive, but we want to be able to treat as many people as possible. Problems only promise to get more severe as the baby boom generation ages and develops more costly chronic diseases. The expensive nature of acute, end-of-life medical care promises even further cost increases.

The theories of innovation suggest one of the root causes of the health care crisis is a focus on the wrong kind of quality. There certainly are circumstances in which patients want and need the most highly trained experts, in which they will bear any cost to be diagnosed and treated. But there are other circumstances as well, circumstances in which patients define quality differently because they want convenience or low prices. In the pursuit of performance, today's one-size-fits-all system does a poor job of meeting these needs. Disruptive innovations can help address these problems. Disruption can create a health care system that treats the hardest cases well and provides convenient, low-cost solutions to reach nonconsumers and to better meet the needs of overshot customers.

Our theories see change. The continual rise of spending on health care creates an unprecedented pot of gold that has attracted some of the world's leading thinkers to capitalize on the opportunity. Science is knocking on the door of ability and, if history is any guide, the door will *be knocked down. Successful disruptive innovations will ultimately move treatment to prevention and will allow each of us to better manage our own health. There are enormous opportunities for both profit and increases in social welfare as innovators help consumers make trade-offs between cost, quality, and access.*

The health care chapter has three general lessons:

- *Scientific progress leads to better categorization and the development of rules guiding prevention and treatment.*
- *Those rules open the door for less-skilled people to do what previously required deep expertise.*
- *Nonmarket forces affect the market for innovation by influencing industry players' motivation and ability.*

Notes

1. Many of the concepts in this chapter were first presented in Clayton M. Christensen, Richard Bohmer, and John Kenagy, "Will Disruptive Innovations Cure Health Care?" *Harvard Business Review*, September–October 2000, 102–117. We also would like to thank Matt Eyring, Dr. Jerome Grossman, Marie Mackey, Chirag Shah, and Tony Ulwick for contributing to this chapter.

2. Employees tend to bear most of the health care cost burden. This is an interesting historical accident. The government froze wages in the 1940s to try to constrain inflation. To try to attract good workers, companies cleverly offered them health benefits, which the Internal Revenue Service decided would be nontaxable. Although the advent of managed care in the 1980s helped to trim fat out of the health care system and to lower costs, costs started rising sharply—on the order of three to five times inflation—in the late 1990s. According to the Kaiser Family Foundation (a nonprofit organization that studies health care), the cost of employer-sponsored health plans (which cover 60 percent of all Americans) increased by 13.9 percent, 12.9 percent, and 10.9 percent in 2003, 2002, and 2001 respectively. These figures come from the Kaiser Family Foundation's annual survey of employer health benefits. For more information, see <http://www.kff.org/insurance/ehbs2003-abstract.cfm> (accessed 25 January 2004).

3. We would like to thank Marie Mackey for her contribution to this section. Specific sources include Carl Berke, interviewed by Marie Mackey, Boston, MA, 26 March 2002; Robin Elise Weiss, "The Rabbit Died!: The History of the Pregnancy Test from Rabbits to hCG," <http://pregnancy.about.com/library/weekly/aa090901a.htm> (accessed 15 April 2002); Rebecca Lipsitz, "Diagnosis at Home: Pregnancy Tests," *Scientific American*, November 2000, 110–111.

4. The RIA test for pregnancy verification was adapted from a general assay that Rosalyn Yalow developed in the 1950s and 1960s. Yalow won a Nobel Prize for this development.

5. The technology that enabled the further simplification of the pregnancy test and its improved sensitivity and accuracy was the discovery of monoclonal antibodies in the 1980s by the French organic chemist Jacques Saussure, who was awarded a Nobel Prize for his work. Before Saussure's landmark work and its incorporation into pregnancy tests, the tests used biological antibodies from animals.

Biological antibodies are polyclonal, meaning that they have multiple binding sites, which introduces the possibility of error in the test results, because they can potentially bind to molecules in the urine sample at sites other than the one specific to hCG.

6. Michael Johnsen, "The Male-Dominated Condom Set Makes Room for the Ladies," *Drug Store News* 24, no. 10 (2002): 19.

7. Christensen et al., "Will Disruptive Innovations Cure Health Care?"

8. Figures based on unpublished statistics obtained from the American Heart Association. Figures do not include outpatient procedures, which could lead to the angioplasty figure being severely undercounted. General figures available in American Heart Association, "Heart and Stroke Statistical Update," <http://www.americanheart.org/downloadable/heart/10148328094661013190990123HS_State_02.pdf> (accessed 16 September 2002).

9. The notion of different levels of providers typically causes some consternation. Nonetheless, historically it required more training to be a generalist than a nurse, more training to be a specialist than a generalist, and more training to be a subspecialist or surgeon than a specialist. In general, fees for service reflected the different levels of training.

10. Interestingly, computer proponents voiced this same argument forty to fifty years ago. They forecast correctly that computers would provide great benefits to society even as they eliminated the need for millions of jobs along the way. Most of the displaced jobholders became free to pursue higher value-added activities as productivity, economic growth, and living standards improved.

Over the long term, of course, wide-scale disruption would lead to the need for a long-term rebalancing of specialists. Hospitals could take proactive measures to make this rebalancing less painful by ensuring that young, highly trained specialists get training in other disciplines.

11. Although these conditions are relatively minor, they make up a large portion of physician office visits. For example, in 2000 the Centers for Disease Control reported more than 31 million doctor visits for sinus-related problems, more than 16 million visits for earaches, and more than 9 million visits for sore throats. Figures from the Centers for Disease Control, <http://www.cdc.gov/nchs/fastats/pdf/ad328.t12.pdf> (accessed 12 July 2002).

12. U.S. Congress, Office of Technology Assessment, "Nurse Practitioners, Physician Assistants and Certified Nurse Midwives: A Policy Analysis," Health Technology Case Study 37, OTA-HCS-37 (Washington, DC: U.S. Government Printing Office, December 1986).

13. We did detailed research to identify the jobs people were looking to get done in health care. The research identified seventy-two unique jobs, including "Know with certainty that body is free from disease, infections, and tumors" and "Maintain concentration and focus." See Anthony W. Ulwick, Clayton M. Christensen, and Jerome H. Grossman, "A Prescription for Health Care Cost Reform," *Strategy & Innovation*, March 2003, 12–13.

14. Interestingly, when most people think of firms driving disruption in health care, they naturally think of device and drug companies. However, many innovations come from practitioners. Innovations should conform to these principles, regardless of where they came from.

15. For more information on Sonosite, see the following sources: Jeremy Dann, "Imaging Innovation: The Portable Ultrasound Market Illustrates the Art and Science of Targeting the 'Low End,'" *Strategy & Innovation*, March 2003, 14–15; Clayton Christensen and Jeremy Dann, "Sonosite—A View Inside," Case 9-602-056 (Boston: Harvard Business School, 2001).

16. Chirag Shah, a 2003 Harvard Business School M.B.A. graduate, noted this during a student research project. His findings are in an unpublished working paper: Chirag D. Shah, Troyen A. Brennan, and C. M. Christensen, "Interventional Radiology: Disrupting Invasive Medicine," 2003. Copies available on request.

17. This illustrates a more general point—outsiders often ascribe too much value to industry expertise and too little value to expertise in the problems managers will confront. The question is not, "Should we get an outsider?" It is, "What kind of outsider should we get?"

18. A 2002 survey showed that almost half of online users visit health-related sites at least monthly. Sites such as WedMD and yahoohealth.com receive more hits than any other category of Internet sites. Jupiter/The NPD Group, Inc., "U.S. Online Activities, 2002, Individual User Survey," May 2002, <http://www.jup.com/sps/data/jupdata.jsp?doc=dl2646> (accessed 16 Septenber 2002).

19. Dr. Mike Magee, "The Evolution of the Patient-Physician Relationship in the United States: Emancipation, Empowerment and Engagement," 13 May 2002, <http://www.positiveprofiles.com/about_us/evolution.htm> (accessed 10 May 2003). In the paper, Dr. Magee refers to a study that found 44 percent of patients thought their relationship with the doctor was a partnership and only 17 percent thought the doctor was a "paternalistic, authoritarian figure."

CHAPTER NINE

INNOVATION OVERSEAS

Using Theory to Assess Corporate and Country Strategies

What insights do the theories of innovation provide about a country's macroeconomic strategy? What can the government do at a broad level to encourage disruptive innovation? What policies stymie disruptive innovation? How should companies think about expanding overseas? Are there ways for them to think about innovation overseas disruptively? What does nonconsumption look like in the international context?

The previous four chapters showed how the theories of innovation help bring clarity to the future of very different industries. In our penultimate chapter we take a slight detour to show how the same theories can provide insight into important issues that cut across industries. We have found the theories to be a useful tool to think about many topics, such as branding and how to organize research and development activities.[1] This chapter specifically shows how to use the theories of innovation to assess a country's macroeconomic environment and a company's internationalization strategy.

There are two main takeaways from this chapter:

1. All else being equal, countries whose economic systems facilitate and motivate disruption have better long-term growth prospects.

2. We should expect companies that follow the principles of disruption when they expand overseas to have high growth potential. Finding simple ways to reach nonconsumers in developing markets provides tremendous room for growth.

Evaluating the Macroeconomic Environment: Seeking the Wheel of Disruption

Assessing the future of an individual firm requires understanding the macroeconomic environments of the countries within which the firm operates. After all, if a country that constitutes a high portion of a firm's sales plunges into economic crisis, it can have dramatic effects on the firm. All else being equal, we would expect firms to be better off seeking opportunities in countries poised for growth rather than in countries caught in cycles of diminishing output, rising prices, and high unemployment. There are a number of useful concepts, such as basic macroeconomic analysis, cluster theory, and comparative advantage, to guide the analysis of a country's strategy.[2] Country analysts should add one more tool to their arsenals by assessing whether a country has an environment that favors wave after wave of disruption. This section explains how to identify environments that encourage this *wheel of disruption* and explains what policymakers can do to increase the probability that a country will avoid economic crisis.[3]

Many of history's best economic and business thinkers have spent their careers understanding what strategies a country can take to successfully increase growth, productivity, and employment. Yet puzzles remain. Consider Japan. From the 1960s to the 1980s, experts praised Japan's high savings rate, lifetime employment, powerful *keiretsus* (loose conglomerations of companies organized around a single bank), and prescient government involvement through the Ministry of International Trade and Industry (MITI). These factors, experts held, drove Japan's miracle growth. High savings generated funds for investment. Lifetime employment let firms act in their best long-term interests rather than worry about short-term market whims. Interlocking networks of companies into *keiretsus* let individual companies take advantage of economies of scope. Government involvement helped champions successfully enter other markets. Many observed these phenomena and heralded Japanese-style management as the wave of the future.

In the 1990s, experts pilloried Japan's high savings rate, lifetime employment, powerful *keiretsus*, and government involvement. These factors, experts held, contributed to Japan's economic malaise. High savings meant consumers weren't consuming. Lifetime employment led to firms being sluggish and reactionary. Interlocked *keiretsus* made change difficult. Government involvement only made matters worse. Japanese-style management seemed a failed experiment that no country would ever seek to revive.

How could a country get it so right and then so wrong? The lens of the disruptive innovation theory provides some answers.

Think about the firms that led Japan's economic miracle: Toyota, Sharp, Matsushita, Sony, Nippon Steel, Kobe Steel, and Canon. *All* of these firms grew to greatness on the backs of disruptive innovations. Toyota established a disruptive foothold with its low-cost Corona model. Sony launched many distinct disruptive innovations such as the transistor radio. Nippon and Kobe Steel started exporting inexpensive steel targeted at the lowest-quality tiers of the American market. Canon created a tabletop copier that brought photocopying to the masses. And so on.

Once these companies established their disruptive footholds, they needed massive investment to march up their sustaining trajectories. Their strategies were clear. Lifetime employees helped them maintain the necessary long-term focus. The government helped them enter foreign markets. Rounds of innovation led "made in Japan" to become a source of pride rather than an epithet. Sony, Sharp, and Matsushita became some of the world's most respected makers of high-quality consumer electronics products. These companies all reached the most demanding customers. They all became some of the biggest firms in the world.

Like many leading companies, these firms ran into the innovator's dilemma. The new disruptive market opportunities that promised the next round of massive growth looked too small and too ill-defined to be interesting. These companies' well-honed, meticulous planning processes enabled them to compete in established markets but did not help them aggressively create new markets. Mergers among ailing corporate behemoths brought stability, but at a cost. Opportunities that looked small and uninteresting to several large companies looked even smaller and even more uninteresting to massively merged companies. MITI's focus over the past fifteen years has been in classic sustaining markets—such as fifth-generation computers and high-definition televisions—that are amenable to rigorous planning. Central planning can help a company or an industry introduce sustaining innovations to move along an established technological trajectory, but it is of little help when searching for the next disruptive niche.[4] In other words, MITI could articulate deliberate plans once the disruptors had taken root and existing but not-yet-served market tiers were obvious and measurable. But deliberate planning before companies established footholds led to investments in overshooting technology.

Compare Japan's experience to that of the United States. The United States's economy certainly hit a speed bump at the beginning of the twenty-first century. Yet those troubles came at the end of two decades of unprecedented prosperity. No one believes the game is over, that the United States is a failed economic model. At one level, U.S. companies have generally followed the same basic pattern as Japanese companies. Most companies have created markets only once. Their focus then turns to exploiting the opportunity by moving aggressively up-market. The ride up is exhilarating and rewarding. Things look great while it lasts. But growth eventually stalls and those companies find the next generation of potentially disruptive opportunities uninteresting. Then what happens? In the United States, the managers leave, take disruptive technology and ideas with them, pick up some funding, and launch their own disruptive attacks. Or entrepreneurs sense an opportunity, raise capital, and launch disruptive attacks. When the disruptor becomes the incumbent and growth stops, what happens? In the U.S. economy, the cycle continues, again and again.

Silicon Valley companies are legendary for spawning their ultimate competitors. Leading semiconductor companies such as Intel were founded by managers who left industry pioneer Fairchild Semiconductors in the 1960s. Managers also left existing semiconductor companies to create next-generation providers such as Xilinx and Altera. Disgruntled managers left networking giant Cisco to form Juniper Networks. The phenomenon occurs outside California as well. Michael Bloomberg developed his disruptive terminal after leaving Salomon Brothers. Industries without such patricide still feature wave after wave of disruption. Think about the retail industry. In each successive generation of disruptive innovation, leaders get toppled, consumers win, and the process begins anew.

This wheel of disruption—firms establishing disruptive footholds, growing, seeing growth stall, squashing internal ideas, and having man-

FIGURE 9-1

The Wheel of Disruption

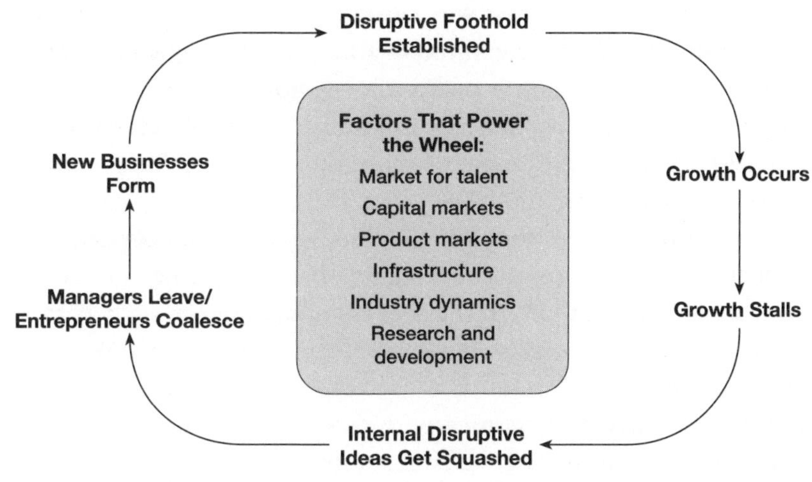

agers leave or entrepreneurs coalesce to get funding and form new businesses to establish new disruptive footholds—is a core microeconomic engine of macroeconomic growth (displayed in figure 9-1). Many of the same writers who suggested that American management was the way of the past and Japanese management the way of the future in the 1980s ascribed much of the United States's subsequent growth to American management. However, we cannot credit the economic success of the United States to the ascendance of American management any more than we can credit the economic struggles of Japan to the descent of Japanese management. The United States has been able to continue moving along the wheel of disruption.[5] Japan has not.

Our research suggests that six factors encourage the wheel of disruption:

1. ***A market for talent*** that is flexible, encourages entrepreneurialism and risk taking, and enables mobility between firms.
2. ***Capital markets*** that help new firms start and grow while targeting disruptive opportunities. Capital market policies that encourage debt financing inhibit the wheel of disruption. Remember, disruptive opportunities start out small and unpredictable.

Bank financing requires high degrees of predictability and steady cash flows. Bank lenders simply lack the flexibility to tolerate the experimental, improvisational manner in which disruptive firms grope their way through the fog of new markets.

3. ***Unconstrained product markets*** that provide ample motivation and ability (particularly access to overshot customers or nonconsumers). One particularly important component is the capacity to find or create new distribution channels that support a disruptive business model.
4. ***A supporting infrastructure*** that has appropriate tax policies, encourages company formation, and has intermediaries that provide "lubrication" to the process of disruption, such as training and education, market research, and verification and accreditation services.
5. ***Vibrant industry dynamics*** with market-based interactions and competition to spur new business models.
6. ***A research and development*** environment that protects intellectual property while directing research toward breaking tradeoffs and applying technology into new markets.

These factors all interact to define an environment that is conducive to disruptive innovation. For example, immobility in Japanese labor markets shackles internal managers who have disruptive ideas. Because an idea and an entrepreneur typically are coupled, when managers can't leave, there aren't any entrepreneurs with ideas seeking to raise funding. This means there is no risk capital infrastructure to fund potentially disruptive ventures. The lack of new disruptive ventures limits growth.

Generally, our models predict that the best long-term growth potential comes from creating an environment that helps develop countrywide processes that support disruptive growth. Attempting to follow a low-end approach simply by arbitraging differences in resource costs (for example, low wage rates) is a flimsy strategy. Processes that stretch across firm boundaries can create defensible, countrywide competitive advantage. Our belief is that between-firm processes are one of the elements that make industry "clusters" a powerful source of competitive advantage. Most countries have chosen to have industry-based clusters; our belief is that a cluster of firms grouped around new-market disruption could be a real growth engine.

What implications does this have for countries beyond the United States and Japan? Singapore had an extraordinary run of success building competitive industry clusters that now produce some of the best products in the world. But in the late 1990s, growth slowed. In 2001, Singapore's gross domestic product (GDP) shrank. Despite a rebound in 2002, Singapore might have a problem. Singaporean companies had reached the high end of their market. To improve living standards, the country needs the next wave of growth to come from disruptive companies.

Singapore can look to Taiwan, a country whose economy exudes Schumpeterian capitalism. Few Taiwanese companies try to muster an all-out attack on powerful global competitors. Instead, thousands of new companies start every year, many with strategies targeting disruptive markets. For example, Taiwan Semiconductor (discussed briefly in chapter 7) avoided a head-to-head battle with leading companies such as Intel by initially manufacturing relatively simple microprocessors. Quanta—a contract manufacturer—started simply. It did bare-bones contract manufacturing of laptops. Little by little, it moved up the industry's value chain. By 2003, it managed the entire production process while doing design work for companies such as Dell and Apple.[6] Although Taiwan has many of the right pieces in place, it too remains vulnerable because it has not created a new-market disruptive engine.

What about South Korea? South Korea's growth largely came by forming large *chaebol* (conglomerates of companies in diverse industries clustered around a parent company) that took advantage of the country's relatively low labor costs and government subsidies to muscle into big markets such as shipbuilding, steel, automobiles, consumer electronics, and computer memory. Korean companies such as Hyundai, Samsung, and Daewoo have had an extraordinary run of success mobilizing their resources to attack sophisticated global competitors. They have not reached the high end of the market and still have room to grow. But the challenge seems clear. As Korean companies grow larger and larger and as labor cost advantages begin to dissipate and shift to other countries, Korea must develop the next round of disruptive attackers. Its reform efforts seem to be finding some success, particularly in encouraging innovation in communications-related industries.

These concepts can help assess developing countries as well. For example, China's recent growth has largely come from foreign direct investment that took advantage of its relatively low labor costs. India, on the other hand, has created firms such as Dr. Reddy's Lab, Wipro, and Infosys, all of which seem to have legitimately disruptive business models. Our theories would predict that India will have greater long-term growth potential if it continues to have an environment that allows such firms to blossom.[7]

Of course, if firms in any of these countries understand the lessons from our work and develop a process to launch disruption after disruption, many factors that were historically important would become irrelevant. For example, Korea could benefit from disruption even if there are environmental factors (such as a lack of labor market mobility, a paucity of start-up capital, and regulations that make it difficult to start new businesses) that inhibit the creation of new firms if the *chaebol* develop the capability to create disruptive growth businesses. Regulations that affect specific product markets and firm-level decisions would then take on much greater importance, because those regulations would impact the operating decisions of those companies.

Indeed, global companies that learn to apply the principles of disruptive innovation in their decisions to create growth markets by expanding overseas could demonstrably change things for the better—not just for themselves but for many of the world's poor. The next section shows the opportunity companies can create if they leap to the bottom of the world's income pyramid.

Evaluating a Firm's International Strategy: Seeking the Great Leap Downward

Companies hoping to create new-growth businesses can choose from three fundamentally different market entry strategies.[8] The first is to develop better products than existing competitors and use them to steal away the competitors' best customers. The second is to create a low-cost business model to launch a low-end disruption targeting price-sensitive customers. The third is to reach nonconsumers. After firms establish themselves in their home markets, they face similar choices if they seek to go abroad. And most of the time the question is when, not if. Firms must grow. Countries are of finite size. When companies tap into all of the growth opportunities within their home countries, they *must* either create new markets domestically or look abroad to satisfy their hunger for growth. Outsiders assessing a firm's international growth strategy should look to see if companies are following the principles of disruptive innovation when they head overseas.

For the sake of simplicity, let's consider an American company called (creatively) Amcom that is formulating an international strategy. What choices does it face? First, it must decide the country or countries on which to focus. Should it try to sell to customers in wealthy countries, such as France, Korea, and Australia? Should it try to sell to customers in middle-income countries such as China, Chile, and Thailand? Or should it seek customers in low-income countries, such as India, Kenya, and Burundi?[9]

Once Amcom picks its target countries, it then must decide which specific customers to go after in that country. Generically speaking, should it target the wealthy, the middle-class, or the poor?

The concept of resources, processes, and values helps to explain most firms' default decisions. What strategy would fit Amcom's values? The natural choice is to find customers that look like existing customers in countries that look like the home market. So Amcom sensibly would look to relatively rich countries. If Amcom did try to go to lower-income countries, it would most likely look for the "upper part of down" by targeting the high-performance products that generate attractive profits in their home markets toward the very wealthiest customers in these low-income countries as well. We can consider these approaches to be sustaining strategies. Companies have a pretty good track record of creating modest growth by using their superior resources to out-muscle local leaders on the sustaining trajectory. However, these sorts of sustaining strategies rarely create major new waves of growth.

For example, General Motors clearly decided to target consumption when it built a Buick plant in China in the late 1990s. This pits the Buick against prestige brands such as Mercedes, BMW, and Lexus to win the hearts and wallets of China's very wealthiest consumers. This effort seems to be wringing out modest profits but has not spurred significant growth. Trying to win this battle has been expensive. Even if GM ultimately does win, it might face difficulties in finding export markets for the Chinese-produced cars. China lacks a suitable infrastructure of materials and suppliers to produce a car that has the functionality and consistency that Western European and North American consumers demand. The list of U.S. companies taking this approach—hoping to somehow find large numbers of affluent customers in developing countries—and struggling is long.[10]

Targeting the Bottom of the Bottom

We have written extensively about the power of a strategy based around disruption. The list of companies that started with legitimate disruptive innovations pretty much maps the list of stocks we wish we owned. Disruptive innovations—especially new-market disruptions—are powerful because they create new growth. If Japanese and American companies have created massive amounts of growth by going down to the "bottom of the top" of the population pyramid depicted in figure 9-2, consider the upside potential of a firm that finds a way to reach nonconsumers close to the pyramid's base.

Nonconsumption is rampant at the bottom of the pyramid. Establishing a foothold here gives a disruptive upside of close to 6 billion people. Consumers in some of these markets could be very happy with very little. Figure 9-2 displays this great leap downward. The figure shows

FIGURE 9-2

The Great Leap Downward

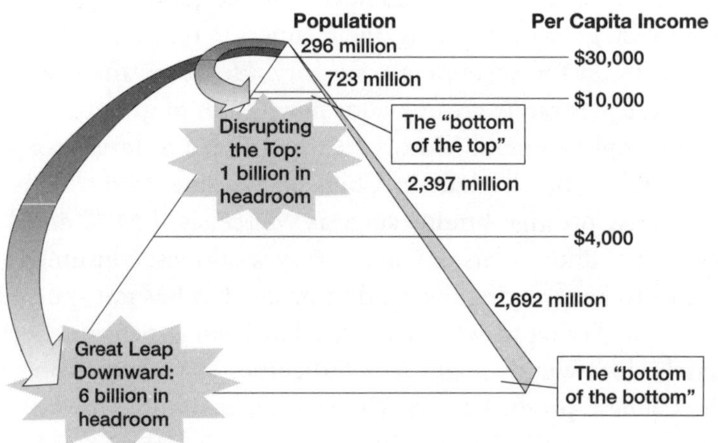

Source: Author analysis, World Bank; per capita income based on purchasing power parities.

how countries with relatively high per capita income (greater than $10,000) have roughly 1 billion people, whereas countries with low per capita income (less than $4,000) have more than 2.5 billion people. The further toward the bottom companies go, the larger the number of nonconsumers and the greater the upside.[11]

How could a company follow this approach? We will explore three specific examples: GM's effort to create minivehicles in China; a Chinese manufacturer's efforts to create a disruptive microwave oven business; and a Bangladeshi company's efforts to develop a telecommunications business. Each effort goes further and further down the income pyramid—GM targets the middle of the middle, Galanz (the Chinese manufacturer) targets the bottom of the middle, and Grameen (the Bangladeshi concern) targets the bottom of the bottom. If successful, all three companies could have tremendous growth potential.

GM and Minivehicles

In 2002, GM announced a joint venture with a Chinese company to produce and sell minivehicles. In contrast to its efforts to target the top of the middle part of the pyramid with its Buick plant, GM's minivehicle program would seem to have more disruptive potential. Minivehicles are small, simple, fuel-efficient, and relatively inexpensive (about $3,000 each). They could prove attractive to the Chinese middle class who historically could not afford automobile ownership. These people are nonconsumers who can be delighted with a relatively simple product.

This strategy of targeting nonconsumers in the Chinese market would seem to have more growth potential than GM's efforts to push its luxury brands. If GM succeeds in creating a business model and a product platform that can profitably make and sell $3,000 vehicles in China, it could find the business to be an ideal platform from which to move up-market. It could build slightly bigger, slightly more powerful cars to sell to other emerging markets such as Thailand, then Russia and Eastern Europe and Latin America. Subsequent product generations with an $8,000 price tag could crack the U.S. used car market. Ultimately, the venture could begin competing in the North American new car market. Targeting this market directly from the beginning would likely be suicidal for GM's China operation because of the demanding performance and reliability standards. And it would be crazy, because there is so much growth for GM to capture along the disruptive path.

This is one reason why leaping down can be so powerful. Finding an economical way to create low-cost products creates a business model that can be a real strategic asset. The business model serves as a platform from which a company can disrupt competitors in other markets.

Galanz and Microwave Ovens

Whether GM can be successful with the challenging opportunity will not become clear for years. But another development in China already shows the potential inherent in reaching further down the income pyramid. A Chinese company called Galanz created a disruptive microwave oven that first took root among nonconsuming Chinese and grew to own a significant share of the world market.[12]

In 1992, Galanz was a textile and garment manufacturer seeking a new growth engine. Microwaves would hardly seem to be a logical choice. The market was mature and shrinking. Most potential consumers in developed markets had microwaves. Products were more than good enough to do what people wanted them to do.

Galanz could have followed a low-end disruptive strategy by using China's relatively low labor costs to build a low-cost microwave oven for export into established markets. This might have worked for a time, until the world's microwave manufacturing operations fully migrated to China—at which point a low-cost strategy would become an equal-cost strategy. So Galanz decided to try to reach nonconsumers domestically. Only 2 percent of all Chinese households owned a microwave oven. Most families just didn't have kitchens large enough to accommodate microwave ovens built with Western kitchens in mind. Galanz's founder Qingde Liang decided to develop a simple, energy-efficient microwave that was small and cheap. Although limited, it certainly trumped nonconsumption.

Galanz started small. Its domestic market share was a meager 2 percent in 1993. But sales steadily increased. Liang took advantage of economies of scale to reduce the product's price and reach a growing number of Chinese consumers. By 2000, it owned the Chinese market with a 76 percent market share. Once Galanz created a business model that could be profitable at low price points, investments to add features and functionality to its product—first for higher-end Chinese customers and then for established customers in developed countries—loomed as very profitable and attractive. Galanz moved steadily up-market to build larger microwaves with more features, beginning to disrupt microwave oven markets in developed countries. By 2002, its global market share was 35 percent.

In 2000, Galanz began trying to replicate its success by disrupting the home air conditioning industry with a low-end product. Again, it built a simple, low-cost, energy-efficient product that was good enough to cool the small homes and apartments in which most Chinese live. To these nonconsumers, even a relatively limited product seems like a gift from above. If Galanz is able to build a business model and product platform that can generate profit at the low price points required to enable middle-income nonconsumers in China to own air conditioners, it could have another disruptive growth business on its hands. Subsequently attacking markets in higher-income countries will be relatively easy.

Grameen and Telecommunications

What about leaping to the very bottom of the pyramid? Are there ways to cleverly target nonconsumers among the 3 billion people who earn less than $2 a day?[13] The Grameen family of businesses in Bangladesh has spawned many potentially disruptive spin-offs.[14] One pertinent example is Grameen Telecom, which brings information and communication technology to rural Bangladesh.

While most of us in the developed world take our phone service for granted, almost half of the world lacks reliable telecommunications service. In some areas, the cost to build or extend a wireline infrastructure is prohibitive. It would require charging people close to their annual incomes to make telephone calls. Enter disruption.

Where most firms would see a hopeless situation, Grameen Telecom saw an opportunity. Its approach? Bring wireless service to the country's poor. It loaned up to $175 each to women who became known as the "wireless women of Grameen." The loan covered the mobile phone's cost, training, and a small solar recharging unit. The women could then sell phone usage on a per-call basis at an affordable price to other villagers.

Users love the service. It is convenient and cheap. It allows farmers to get vital information such as crop prices without traveling great distances. They can place orders with distributors without relying on the postal system. Estimates peg per-call savings of up to $10, almost 10 percent of household monthly income. The program has other benefits as well. People travel less, saving wear and tear on shoddy roads. Phone operators earn $300 a year. They plow this money back into education and health care, increasing the country's human capital. Their status in the village increases considerably.

So operators are happy and users are happy. What about Grameen? Its business model allows it to be *very* profitable. If extended to all of Bangladesh, the business could generate more than $100 million a year. And just think of the opportunities to bring the model to rural areas in other developing countries such as China and India. This is how companies build new-growth markets—by competing against nonconsumption. The lower in the pyramid you start, the greater the up-market potential.

Another Approach: Fueling Local Disruption

A company can also follow a leap to the bottom strategy by finding a way to enable the formation of local disruptive firms. The company could provide the fuel that lets firms emerge to target local nonconsumers.

Consider the experience of a chemical company we recently worked with. The company sold a chemical to a global maker of personal care products, such as shampoos, skin lotions, and oral care products. The chemical company lived a tough life. Its customer viewed the chemical as a commodity. And indeed, in these applications it was the personal care products company that did the hard work. It used the chemical as part of its technologically sophisticated, proprietary product formulation.

The chemical company was searching for a strategy to enter the Chinese market. One approach would be to continue as a supplier to the consumer products company—but that would continue to consign the chemical company to a dismal life as a commodity supplier. Another approach, however, would be to use the chemical to build proprietary building blocks that made it foolproof for Chinese companies with little technical know-how to formulate shampoo and other personal care products. Without huge advertising budgets, these local companies could make money at price points unimaginable to the consumer products company. Their product wouldn't be good enough for Western consumers, but it would be better than nothing at all. Being the fertilizer for thousands of blooming flowers is a good thing. This path is not easy. It requires some forward integration, work to build a new market, and the ability to think about the consumer in a very different manner. But a company that follows this approach could have a nice business on its hands.

How Advanced Technologies Could Benefit from Leaping Down

The approach outlined in this chapter also has implications for how companies commercialize the advanced technologies they develop in their laboratories. The best commercialization path might not be through high-income countries. In fact, delighting nonconsumers at the bottom of the pyramid might be the way companies hone critical skills in commercializing new technologies. This approach has significant social benefits as well, bringing valuable services to populations traditionally ignored by leading companies.

Leaping downward can also help companies market environmentally friendly products. Western companies often struggle to market "green" solutions. In most developed markets, customers are not willing to sacrifice performance and functionality in exchange for environmentally friendly products, especially when the products are more expensive. Competing against consumption with a not-good-enough product is always a tall order. Early-generation technologies are just not up to the task. But when consumers compare a green technology to nothing at all, its limitations often aren't a roadblock at all.

Think about the use of photovoltaics (solar energy). Making the technology as good as the electrical power most of us receive from a central generation plant is a daunting proposition. Sometimes there is just not enough sunlight to power all of the appliances in a modern business or home. And customers used to ubiquitous, cheap electricity just will not put up with random "cloudouts." But what about a rural village in India? The villagers really don't care if solar technology doesn't work all of the time. It certainly is better than no electricity at all. Delivering solar energy could appear cheap, because delivery doesn't require building an expensive infrastructure with centralized generators and an organized power grid. Reaching this market doesn't require creating backup facilities or making photovoltaics incredibly powerful, as is the case in North American and Europe. Existing technology isn't yet good enough to satisfy customers in mainstream markets. But companies can fabricate much less expensive solar cells to target homes that have never had electricity. In this context, enough electricity to power a lightbulb would delight customers. As participating companies produce increasing quantities of photovoltaics, they will improve their products, enabling them to move up and across markets.

What to Watch for in a Leap Downward

What would we see that would lead us to believe an established company might have a chance at making the great leap downward?

We would bet against a company that decided to address the opportunity from within one of its mainstream organizations. The developing-country business would likely inherit the overall corporate cost structure and processes. The company might begin with aggressive growth intentions and spend some money, but it would scuttle the venture when belt-tightening inevitably ensued. The company would say something like, "Despite our best intentions, we just couldn't make it work. The goal is noble. But after all, we are a for-profit company."

This is a clear circumstance in which a company would need to set up a separate organization free to develop its own prioritization criteria. The company should not give the separate organization too much financing, because the key to success is the creation of a business model whose cost structure enables profits to be earned at low price points. The organization would have to develop its own processes. Research shows companies entering developing markets must cultivate long-term relationships with partners such as local communities, local companies, and nongovernmental organizations.[15]

As companies grow, they tend to seek markets outside of their home geographies—particularly if they come from small countries where domestic growth is limited. Understanding whether these efforts will be successful requires analyzing the environment in the new country and the firm's globalization strategy. This chapter shows how the theories of innovation can help assess both sets of issues.

Countries that encourage the wheel of disruption have the best long-term growth prospects. Countries that lack such a wheel more often than not will find themselves trapped in a country-wide innovator's dilemma. The market for talent, capital markets, product markets, infrastructure, industry dynamics, and R&D environment all influence the wheel of disruption.

A firm can have a long successful ride globalizing in a sustaining fashion—finding similar-looking customers in similar-looking markets and selling them premium products for premium prices.

But ultimately they run out of room. Companies then often fall off the tracks by taking high-cost products to nonconsumers in lower-income markets.

Another option is to establish a presence in a foreign market in disruption. Disruptive strategies promise tremendous room to move up-market. And if a company takes root in a disruptive new market in a low-income country, it has unprecedented *room to move up-market. To attack these opportunities, companies need to use different evaluation criteria and build different capabilities.*

Embracing the principles of disruptive innovation—starting small and simple and helping people accomplish the things they care deeply about—is a fundamental engine of economic growth and development. It is the mechanism that transformed countries such as Japan, Korea, Taiwan, and Singapore into economically prosperous nations. And it is the strategy that enables firms such as Toyota, Samsung, and Taiwan Semiconductor to become huge global leaders in their industries.

Our intent is not to encourage firms to set out to improve developing countries. Most people agree that improving the lives of the mass of people who have very little is a noble goal. However, direct aid and Western-style reform have proven largely ineffective. We believe that companies and countries that follow disruptive principles will end up having significantly more positive impact than aid and reform efforts have had.

Notes

1. The disruptive innovation model can help explain when and where companies can create value through brands. In short, we suggest there are three different types of brands: performance brands, convenience brands, and price brands. Overshooting—providing more performance than customers can use—leads to performance brands migrating from product producers to component providers. Overshooting also creates opportunities for retailers to create convenience and price brands. For more information, see Clayton M. Christensen and Scott D. Anthony, "Performance, Convenience, Price: What's Your Brand About," *Strategy & Innovation*, November–December 2003, 1–5.

The VCE theory helps explain when companies should outsource product development. Companies should conduct product development internally when the types of improvements the market will value are well understood, product development is expensive, and the product development process relies on interactions with other important internal processes such as manufacturing and distribution. When the types of improvements the market will value are *not* well understood, product development is cheap, and the product development process does *not* rely on interactions with other important internal processes, companies should consider turning product development over to the marketplace. For more information, see Clayton M. Christensen and Scott D. Anthony, "Do You Know What You Do Best?" *Strategy & Innovation*, September–October 2003, 1–5.

The VCE theory can also help companies decide how to optimally organize their research and development activities. When research is aligned to drive improvements at firm-controlled points of interdependence, companies can capture attractive returns. Firms that develop breakthrough materials or components that do not fit into their existing businesses face a long and difficult commercialization path. Furthermore, research focused on points of modularity tends to leak to specialist companies. Companies need to ensure that their research activities match the industry's circumstances. See Clayton M. Christensen and Henry Chesbrough, "Technology Markets, Technology Organization, and Appropriating the Returns of Research," working paper 99-115, Harvard Business School, Boston, 2001.

2. There are of course numerous concepts and works that analysts can use to help understand a country's future. Some classics include: Milton Friedman, *Capitalism and Freedom* (Chicago: University of Chicago Press, 1962); John Maynard Keynes, *The General Theory of Employment, Interest and Money* (New York: Harcourt, Brace, 1936); Michael E. Porter, *The Competitive Advantage of Nations* (New York: Free Press, 1990); David Ricardo, *On the Principles of Political Economy and Taxation* (Washington, DC: John B. Bell, 1819); Amartya Sen, *Collective Choice and Social Welfare* (San Francisco: Holden-Day, 1970); and Robert M. Solow, "A Contribution to the Theory of Economic Growth," *Quarterly Journal of Economics* 70, no.1 (February 1956): 65–94.

3. Many of these concepts are discussed in greater depth in Clayton Christensen, Craig Thomas, and Stuart Hart, "The Great Disruption," *Foreign Affairs*, March–April 2001, 80–95.

4. As of the writing of this book, Japan has undertaken successive waves of reform to try to address its systemic problems. Hopefully, reformers will create an environment that fosters the next generation of disruptive innovation.

5. There were many structural changes in the U.S. economy during the 1970s and 1980s that helped rev up the wheel of disruption, such as lower marginal tax rates, decreased government regulation, increased government spending, and important technological developments.

6. Bruce Einhorn,"Laptop King: In a Year That's Decimated High Tech, Taiwan's Unstoppable Quanta Is Posting Double-Digit Sales Growth," *BusinessWeek*, 5 November 2001, 48.

7. Harvard Business School professor Tarun Khanna has done some very insightful research and writing on this topic. See, for instance, Yasheng Huang and Tarun Khanna, "Can India Overtake China?" *Foreign Policy*, July–August 2003, 74–81.

8. Many of the concepts in this section are discussed in greater detail in Clayton M. Christensen and Stuart L. Hart, "The Great Leap: Driving Innovation from the Base of the Pyramid," *MIT Sloan Management Review* (Fall 2002): 51–56. University of Michigan Business School professor C. K. Prahalad has studied these issues extensively. Prahalad also concludes that firms need to jump to the bottom of the world's income pyramid and has numerous excellent suggestions for how to do it. See, for instance, C. K. Prahalad and Allen Hammond, "Serving the World's Poor Profitably," *Harvard Business Review*, September 2002, 48–57.

9. These countries are random selections based on their gross national income (GNI) per capita using the World Bank's Atlas method, which takes account of different standards of living in different countries. So-called high-income countries had GNI per capita of $27,680 in 2001; France's was $25,280, Korea's was $18,110, and Australia's was $25,780. Middle-income countries had GNI per capita of $5,710; China's was $4,260, Chile's was $9,420, and Thailand's was $6,550. Low-income countries had GNI per capita of $2,040; India's was $2,450, Kenya's was $1,020, and Burundi's was $590. Data from the World Bank. See "GNI per Capita 2002, Atlas Method and PPP," <http://www.worldbank.org/data/databytopic/GNIPC.pdf> (accessed 19 November 2002).

10. For example, U.S. automakers historically struggled to enter lower-income countries, Home Depot struggled to expand to Latin America, Wal-Mart's early globalization efforts struggled, and Procter & Gamble initially encountered difficulties in Asia. The list goes on.

11. Data comes from online databases accessible via the World Bank. See <http://www.worldbank.org/data/> (accessed 19 November 2002). Of course, a more precise measure would segment by people, not by countries. In other words, all of India (per capita income of about $2,500) gets lumped into the bottom of the pyramid, even though there is a large middle and upper class in India with a per capita income of well more than $4,000 a year.

12. The Galanz case study resulted from an independent student research project done in 2002 by Linda Cai, Harvard Business School class of 2002.

13. World Bank data for 1998 showed about 2.8 billion people living on less than $2 a day. See table 1, "Population Living on Less than $1 per Day and Headcount Index in Developing Countries, 1987, 1990, and 1998," <http://www.worldbank.org/poverty/data/trends/income.htm#table1> (accessed 13 February 2003).

14. A good summary of this approach can be found in D. Richardson, R. Ramirez, and M. Haq, "Grameen Telecom's Village Phone Programme in Rural Bangladesh: A Multi-Media Case Study," (Guelph, Ontario: TeleCommons Development Group, 2000). The Grameen family started with Grameen Bank, one of the first so-called microcredit lenders, more than twenty years ago.

15. The Base of the Pyramid Learning Laboratory at the University of North Carolina's Kenan-Flagler Business School has conducted extensive case study research on multinational corporations and indigenous ventures focused on the base of the pyramid. See <http://www.kenan-flagler.unc.edu/KI/cse/bop.cfm> for more information (accessed 13 February 2003).

CHAPTER TEN

BREAKING THE WIRE

The Future of Telecommunications

How will the widespread deployment of data networks affect companies that provide telecommunications services? Does convergence mean the end of companies such as AT&T and Verizon? Where will specialists win? Where will generalists win? How will cable companies affect the telecommunications industry? Why did companies spend so much money on third-generation wireless licenses? How did NTT DoCoMo's i-Mode wireless service grow so rapidly? Will 802.11 networks continue to grow and improve? Who will capture that growth? What surprising developments could shake up the industry? What do the theories of innovation have to say about the telecommunications industry?

We close *Seeing What's Next* by returning to the telecommunications industry. 2003 was another tough year for the industry. Companies were still struggling to recover from the telecommunications burnout of 2000 to 2002. Underpinning the carnage, however, were important technological shifts that will likely be the platforms for the next wave of growth. Important innovations altered the network core, attempted to unlock the metropolitan area, freed users from wires, and provided users many new devices and service offerings. Data quickly was replacing voice as the dominant form of network traffic. Anything, including voice signals, could be digitized into packets, blasted out into the network, and reassembled at its destination. While many competitive local exchange carriers (CLECs) fell, some remained and slowly gained traction. Long-distance companies felt pain as wireless companies offered seemingly unbeatable rates for the same service. Cable companies emerged as legitimate competitors for local phone service.

In the introduction, we noted how incumbent service providers largely co-opted the first disruptive threat: wireless voice (see "Wireless Voice" for a brief future prognostication). Our specific goal in this chapter is to evaluate whether any of these new developments could pose a credible long-term threat to incumbents.

The short answer is yes. Companies are bringing voice services into new contexts, building low-cost business models, and developing new skills. Potential competitors run the gamut from large cable companies such as Time Warner and Comcast to technology firms such as Cisco and Microsoft to specialist phone providers such as Vonage to nontraditional firms in deceivingly independent markets such as instant messaging.

Disruptive growth is everywhere. But entrant and incumbent choices, as always, will play a large part in determining the future. Our approach highlights the most likely outcome of a potential technology's evolution and identifies the crucial decisions firms could make to change that outcome. The tools we have presented can be used to evaluate the firms that we mention as well as the countless others that will introduce products and services before this book even hits the shelves.

The telecommunications industry is large and complicated. To focus the analysis, in this chapter we evaluate four specific developments:[1]

1. ***Voice over Internet Protocol (VoIP):*** Incumbents have high motivation and the right skills to master VoIP and associated technologies. However, the rise of the Internet protocol (IP) could lead to the decoupling of services from the transport medium, opening the door for specialist providers to take valuable pieces of business away from incumbents.
2. ***Cable telephony:*** Cable companies are largely following sustaining strategies that will result in bruising battles for market share. A disruptive path using IP to reach overshot customers or nonconsuming contexts could be a better long-term path.
3. ***Wireless data:*** Although 802.11 is creating disruptive growth, overlapping value networks and complementary business models

Wireless Voice

The disruption of the wired connection will place traditional businesses under increasing pressure. Wireless companies that do not have wired line assets, such as T-Mobile (a service of Deutsche Telekom) and Nextel, are leading the charge to attract more and more customers away from their wired connections. Over time, service providers such as Verizon who operate both wired line and wireless networks will face uncomfortable decisions about how much they are willing to let their wireless business cannibalize their wired-line voice business. The outcome of this decision will have a direct impact on how local telephone companies plan to continue to support their legacy consumer wired-line business with its complex cross-subsidized price supports. Operators have no option but to allow this cannibalization to occur.

It is important to point out that the result of this disruption does not necessarily obviate a wired line to a business or residence. Increased demand for data and bandwidth is changing the job that wired connections serve. New technologies such as digital subscriber line (DSL) have emerged at the high end of the market. In the short term, the emergence of a new, demanding application will prolong the life and profitability of the wire. Because incumbent service providers have largely co-opted wireless technologies and have a new up-market use for their wire, our assessment is that they are unlikely to be overturned by wireless voice technologies in the near future.

are likely to drive incumbent co-option. Companies making the right choices and the emergence of next-generation technologies such as 802.16 could facilitate disruption.

4. ***Activities in fringe markets:*** Developments in new contexts such as instant messaging could ultimately turn out to be the best bets for long-term disruptive growth.

To make a tough chapter more digestible, we have included short conclusions at the end of each development. Before diving in, however, we must set the stage. We begin by taking the incumbent tale of the tape and looking for the signals of change discussed in chapter 1.

Setting the Stage: Incumbent Tale of the Tape and Industry Circumstances

Before evaluating threats to the incumbents, we need to explore who these incumbents are and discuss the context within which they operate.

Broadly speaking, incumbents in this context are large, traditional telecommunications services providers. In 2004, this included the four Regional Bell Operating Companies, or RBOCs (Verizon, SBC Communications, BellSouth, and Qwest), the three major interexchange carriers, or IXCs (Sprint, MCI, and AT&T), and independent wireless-only providers (such as Nextel, T-Mobile, and AT&T Wireless, which has since been acquired by Cingular Wireless, which is owned by SBC and BellSouth).

How do these companies line up to the tale of the tape? Obvious resources include their cash balances, existing customers, and powerful brands. Incumbents' most powerful resources are their established networks. Unless new entrants want to undergo the substantial costs of building a new network, they have to rely on the incumbents' existing networks. Value networks thus often overlap. This means technologies that rely on physical network connections are often vulnerable to co-option.

Some of incumbents' most important skills include building and maintaining large networks, managing small incremental transactions, handling government regulation, and providing network services. The last skill is particularly important. Communications involves two elements—transport and value-added services. Historically, the two elements required integration: Delivery of telecommunications services inextricably relied on the physical connection medium. Most communications products and services depended on interacting with the physical infrastructure for functions such as origination, circuit switching, provisioning, billing, and termination. Providers had to integrate transport and services to optimize network performance. Their goal was to push reliability toward the so-called five 9s, which means a network works 99.999 percent of the time. That equates to annual downtime of about five minutes.

Historically, integration provided an unassailable source of competitive advantage. You simply couldn't be a specialist provider. Changes to service offerings required changes to the transport medium, and vice versa.

Finally, let's turn to incumbent values. Communications tends to be a high gross-margin business. It has to be. Providers need high marginal contributions to amortize the fixed costs of building and maintaining networks. Providers make money selling local and long-distance transport and services. They sell to residences and businesses. Industry players historically focused on voice calls. They billed and allocated costs based on the length, location, and time of day of a user's call. Over time, providers responded to increasing demands for data communications by overlaying these services onto their existing business models and networks.

In 2003, the seven biggest providers (Verizon, MCI WorldCom, SBC, BellSouth, Qwest, AT&T, and Sprint) had more than $200 billion in combined revenues, with gross margins of close to 60 percent. RBOCs earned slightly more revenue from consumers (about 55 percent); IXCs earned significantly more revenue from businesses (about 60 percent to 70 percent). Data and wireless each made up about 20 percent of revenues.[2]

Incumbent firms' resources, processes, and values define their strengths and their potential weaknesses. Next, we turn from understanding industry players to identifying whether circumstances support the introduction of disruptive innovations.

Signals of Change in Telecommunications

In chapter 1 we suggested that nonconsumers and overshot customers presented the greatest opportunities for industry-reshaping innovations. Could we see signs of these customer groups as of the beginning of 2004?

Are there nonconsumers? At a basic level, this question seems ludicrous—just about every house and every business has multiple telephones. But there is always an important secondary question to ask: Are there ways to bring communications into new contexts where people might *want* to use their voices to communicate but *cannot*? The answer would seem to be yes. People generally cannot use their voices to communicate in many data settings. For example, people collaborating with their colleagues over the Internet might want an easy way to speak with them. Teenagers communicating with friends using instant messaging technology have to type, not speak. There are other types of nonconsumers as well. For example, many small business owners lack the ability to simply and conveniently install world-class phone services. Developing products and services to serve these customers could drive disruptive growth.

The other group that often heralds change is overshot customers. Has the functionality and reliability of the services that existing companies provide overshot what their customers can use? It is a controversial question.

The public switched telephone network (PSTN) that provides telephone service is unquestionably an incredible invention. Voice communication is possible in even the most extreme emergency conditions. Because the PSTN provides power, trees toppled by a blizzard could knock down every power line in your neighborhood and your phone would still have a dial tone.

Yet there are signals that some customers find these capabilities to be more than they need. Customers increasingly use wireless phones instead of landline service. They tolerate relatively poorer-performing technologies in exchange for convenience and what they perceive to be low marginal prices for individual calls.[3] The proliferation of cordless phones, which are effectively useless in a power outage, also raises doubts as to how many households value the lifeline attributes of traditional networks. All of this evidence suggests that new benefits around convenience and portability are trumping functionality and reliability in some market tiers.[4]

In chapter 1 we suggested also assessing the nonmarket context for innovation. The government has been on a twenty-year run of attempting to open markets to competition. Despite these efforts, the residential market remains awash in regulation. Therefore, a natural place to look for emerging innovation is the within-enterprise market, where companies have both the ability and motivation to innovate.

In summary, overshot customers and nonconsuming contexts exist, and the within-enterprise market has historically been the best place to incubate innovation. The next four sections look at developments taking advantage of these conditions to assess their implications for established companies.

1. Voice over Internet Protocol: Co-option with Consequences

Over the last twenty years, companies have built sprawling networks to transmit bits and bytes of data. The rise of data networking is a classic disruptive story—a new-growth market in which entrants such as Cisco Systems, 3COM, and Sun captured most of the growth (see "Cisco and IBM" for a tale of disruption).[5] Data networks led to the creation of new forms of communication, such as e-mail. Generally, data networks are flexible and offer low transmission costs. But they historically have not been good enough for voice traffic.[6]

This section focuses on VoIP, a means to send voice calls over data networks. VoIP could be implemented as a sustaining innovation to lower internal costs and add high-value features and functionality, as a low-end disruption bringing good-enough service at low price points, or as a new-market disruption that makes high-end features more affordable to the masses. VoIP's improvement could spell trouble for traditional equipment providers—in fact, router makers such as Cisco have already disruptively pinned providers of circuit-switching equipment such as Lucent to the extreme high end of the market. But the theories of innovation suggest incumbent service providers will be able and motivated to respond to VoIP. Service providers have skills in building and maintaining networks. They are motivated to go after high-margin revenue opportunities. Overlapping value networks could give them a seemingly easy way to co-opt any potential threats. Incumbent service providers could then choose to use VoIP to create new growth by providing new types of service.

VoIP presents some real challenges, however. If incumbents miss the new growth opportunities, co-option could be cannibalistic. Even worse, incumbents have to contend with the potential decoupling of services from the transport medium—in our language, a widespread modularization where value would migrate to specialists at different points in the value chain. The products and services that VoIP enable are based on business models that are not constrained by time increments and physical connections. Incumbents' resources, processes, and values may not be well suited to succeed in this emerging context.

A Brief History of VoIP: From Hackers to IP/PBXs

Using the any-to-any Internet protocol to transmit voice calls is fundamentally different from using point-to-point, connection-oriented technologies. IP is connectionless. It provides best-effort delivery of packets over a data network. Data is divided into self-contained packets and blasted in every direction along the network.

Although using IP is flexible and cheap, it historically was not good enough for voice transmissions.[7] Some readers might be surprised to

Cisco and IBM

In the 1980s, users realized connecting personal computers made it easier to do things they historically prioritized, such as sharing information with colleagues. They created rudimentary networks with local area network (LAN) technology as the key connector.

One key problem with LANs was that every piece of equipment on a LAN was connected to every other piece of equipment. This worked fine in small settings but was very inefficient as LANs grew to include more users.

Cisco's router solved this problem. The multiprotocol router "spoke" the varied "dialects" (Appletalk, Ethernet, and so on) of the LAN, translated them into the Internet protocol (IP)—the open, connectionless, best-effort protocol—and layered on intelligence that forwarded packets to their ultimate destination. It acted as the traffic cop for the emerging LANs. It was cheap, flexible, and scalable. Easily added to any network, routers enabled companies to interconnect LANs while minimizing network degradation and maximizing network efficiency. Cisco's sales grew dramatically.

Cisco then looked up to the mission-critical data transport market, where IBM and its system network architecture (SNA) dominated and where most of the *real* networking money was made.

At the time, centralized mainframes dominated enterprise computing. Companies reserved these large, expensive machines for tasks critical to day-to-day operations, such as transactions processing. The sprawling private data networks found throughout a major enterprise contained high-quality, point-to-point, circuit-switched connections. Those connections were expensive to install, provision, and maintain.

learn that voice is an extremely demanding application. But consider the "simple" task of completing a voice call. It requires flawless real-time interactions between numerous components at different levels of the telecommunications network. Packets traveling over a connectionless network can be delayed or lost. Those delays or disappearances are fine for e-mail messages. Packets can be continuously retransmitted until the whole message can be displayed. But for voice conversations, lost packets often result in significant call quality degradation.[8]

IBM dominated this market with an overwhelming share of the mainframe, data application development, and data transport markets.

At the core of every system was SNA—a highly reliable, proprietary computing language and networking scheme designed by IBM specifically for its enterprise data business systems.

So Cisco added the capability to transmit SNA to its multiprotocol router. Instead of traveling over dedicated circuits, SNA data traffic could be split into IP packets, routed over a private data network, and reassembled at its destination. It lacked the reliability and security of a point-to-point connection provided by SNA over dedicated leased lines. But it was good enough. Heavy-duty data networking could now be accomplished in a flexible, simpler, and more inexpensive way.

Sales of routers with SNA over IP technology skyrocketed. IBM tried to respond by introducing its own line of internetworking products. But the multiprotocol router had characteristics that made response all but impossible. IBM developed its capabilities in the circuit-switched world. Cisco had developed a unique capability to discover, understand, and incorporate emerging internetworking concepts into its products. IBM's products were inferior to those developed by firms that lived by the new paradigm. Cisco sold to a different type of customer who obviously did not subscribe to the credo "you can't get fired for picking IBM." IBM ceded increasing market share.

In September 1999, IBM arranged to sell its entire internetworking business to Cisco for $2 billion. IBM shifted its focus to what it described as "higher-margin sales support and service opportunities." The 60 percent plus gross margins in its internetworking products that delighted Cisco were incongruent with IBM's values.

Early users didn't care about VoIP's low quality. Because calls traveled over the public network, they were incredibly cheap.[9] Students, immigrants, and others who had tight budgets or placed frequent overseas calls gravitated to the almost-free calls despite quality that others would find unacceptable. Companies such as Net2Phone and DialPad capitalized on the potential low-end disruptive growth opportunity. But prices of international calls dropped, limiting these arbitrage opportunities. (We will discuss VoIP's recent reemergence in the consumer market later.)

At this stage, VoIP was not yet good enough to move into more demanding market tiers. But, as always, data networks got better. Voice quality improved. VoIP next took root within enterprises. Firms' high-speed internal IP networks were the perfect environment to experiment with VoIP. These private, high-bandwidth "walled gardens" minimized quality degradation because data only traveled short distances over "pipes" with ample bandwidth. Companies could use what are known as IP-based private branch exchanges (IP-PBXs) to route all of their internal voice traffic over their data networks (see "Disruptive Equipment?" for an assessment of innovation in the PBX market).

Companies running all of their voice traffic over their data networks could theoretically avoid using existing service providers while making calls between branch offices. But they still would rely on existing providers for all of their other telecommunications services. And, of course, they would still need a data networking provider to build their network infrastructure.

Next, we will look at two potential disruptive paths for VoIP—one growing out of the walled garden and one sprouting up in the consumer market—and discuss the potential implications for incumbent service providers.

VoIP in the Enterprise: Evidence Suggests Co-option

The full-scale disruptive path for VoIP emanating from the enterprise would involve new providers pulling business customers away from incumbents' voice networks to their new data networks. How would this scenario play out?

As the functionality and reliability of VoIP improve and companies implement the technology's convenience-related features, VoIP could extend beyond the walled garden of a single corporation to facilitate collaboration among companies. This improvement trajectory would allow new service providers to develop the skills to offer viable alternatives to incumbent service providers. Building large-scale networks would allow new service providers to solve the interdependencies involved in providing network services (such as network restoration) in a pure IP network, eventually allowing the provision of advanced services such as call center management.[10]

Disruptive Equipment?

IP-PBXs were a key enabler of the emergence of VoIP within enterprises. IP-PBXs are software solutions that run on off-the-shelf servers. IP-PBXs sold by companies such as Cisco, Mitel, and Avaya (a Lucent spinoff) let companies run both voice and data traffic over a single network.

IP-PBXs compete against traditional PBXs, sold by industry heavyweights such as Lucent, Nortel, Alcatel, and Siemens. These systems are expensive and complex. They handle end-to-end digital transmission of voice throughout an enterprise as well as related services such as voice mail and three-way calling.

IP-PBXs created disruptive growth. Traditional PBXs are centralized and proprietary. IP-PBXs are cheaper and less complicated and allow the creation of new integrated services such as unified messaging (the combination of voice, e-mail, and fax). They enable IT managers to do things, such as relocate phones and program new features, that in the past required specialized skill. They move voice services into a data context.

One could chart a disruptive path for IP-PBXs. Traditional PBX manufacturers rely on a large sales force and significant postsales service to maintain the systems once installed. IP-PBXs are more off the shelf, requiring less postsales service. They are growing in smaller businesses that large incumbents cannot adequately serve. Additionally, creating high-performing IP-PBXs requires unique skills honed in the data world. For example, companies with expert knowledge in how data networks function can design their IP-PBXs to interdependently operate with an enterprise's existing data network to achieve reliability that is far higher than an individual system. Traditional PBX manufacturers just don't seem to have those skills.

In early 2004, leading PBX manufacturers were experimenting with hybrid solutions that incorporated IP technology into their existing products and services. These efforts face a low probability of success because they run counter to an incumbent's processes and values.

Who exactly would these new service providers be? They could be companies providing IP-PBX services, such as Cisco and Mitel. A natural up-market move for these companies would be to provide broader communications service. Or they could be new companies building pure data networks, such as Level3. Although Level3's original intent was to provide pure connectivity, layering on services would be a natural up-market move. Alternatively, it could be emerging Internet telephone service providers such as Vonage that have taken root in the consumer and small business market, as we discuss later in the chapter.

If these new providers could offer good-enough services among companies, their next step would be to pull existing business voice services into the data context, pushing incumbents out of the enterprise market. They would then complete the disruption by moving to the residential market.

However, our assessment is that incumbent service providers seem to be both motivated and able to co-opt IP-based technologies and compete in the data context. Why? For one reason, their customers want IP solutions. Furthermore, IP technologies allow service providers to lower their own operating costs. Existing service provider business models feature high gross margins. Using IP will actually enable firms to sustain their business models by lowering their operating costs and increasing their gross margins. This suggests incumbents will be motivated to fight. Furthermore, the symbiotic relationship between data networks and voice networks gives incumbents an easy response mechanism. The two value networks have significant points of overlap.[11] Leading telecommunications service providers are big customers of IP companies, particularly for interconnection and long-haul backbone capacity. As of the writing of this book, some companies such as Level3 were attempting to build their own end-to-end networks. But incumbents such as MCI and AT&T controlled the world's largest data networks. The RBOCs were gearing to expand their data offerings. Finally, some incumbent strengths would come in quite handy in an IP world. For example, the capacity to manage complex networks could be a competitive advantage.

In 2004, evidence indicated co-option. Large customers were demanding that their carriers offer IP-based solutions. Incumbents had begun to deploy and offer IP-based networking solutions. To meet customers' increasing requirements for reliability and still allow them to take advantage of IP-based solutions, operators appeared to be meshing highly reliable dedicated point-to-point connections such as asynchronous transfer mode (ATM), frame relay, and multiprotocol label switching (MPLS) with IP networking solutions, adopting the technologies that incubated in the new plane of competition. The rationale for this decision was largely to blend some of the cost and flexibility advantages of IP any-to-any networking with the security and reliability of the point-to-point technologies.

Done correctly, VoIP could actually be an exciting new-market disruptive growth opportunity for incumbents. For example, VoIP could be used to offer small business owners a simple and affordable means of installing and operating a world-class telephone network. Currently, small businesses tend to do the best they can with limited-functionality phone systems because they have neither the skills nor the financial resources to install the high-end systems that most large corporations take for granted. Making it simpler and more affordable to install and operate a telephone network could provide avenues for growth across this as well as other markets.

What factors could limit the incumbents' abilities to co-opt IP? One big gating factor would be whether incumbents delay the fixed costs of building new networks until IP gets good enough. Waiting might seem sensible, but it could allow entrants to develop a set of asymmetric skills. Another obstacle could be the different mind-set and business model that data requires, one based on capacity rather than minutes of use. Incumbents could attempt to force old paradigms on new technologies. These factors give entrants a chance. But we believe these potential asymmetries are too weak to allow for disruption. Instead a meshed networking world will most likely prevail.

VoIP in the Residence: Looming Low-End Battles

By 2004, specialist VoIP providers had begun to make noise in the residential market. One particularly interesting company was Vonage. Privately held, Vonage targeted customers who purchased high-speed Internet access from their telephone or cable operators. It provided those customers with a small, low-priced box that plugged into their network connections. Consumers would plug a normal telephone into the box and use it to make and receive telephone calls. Vonage would transmit the calls as data packets over the public Internet before converting them to voice signals for call completion.

On many dimensions, Vonage's service was lower quality than the incumbent operator's service. Because calls traveled over the public Internet, customers would occasionally experience call quality problems as packets were lost or delayed. Vonage did not offer line powering, which meant that phone service would be unavailable in a power failure. Vonage also did not offer a cadre of customer service representatives to help with service-related problems, instead relying on a lower-cost combination of limited call center availability and Web-based service. In exchange, however, it offered customers very low prices. Users paid a flat monthly rate to enjoy unlimited local and long distance calls and inexpensive international calls. Additionally, users could plug their Vonage boxes into networks in other locations and receive calls on their "home" phone number.

Vonage has many of the characteristics of a low-end disruption. It offers stripped-down, lower-priced service. Its cost structure, with low overhead costs and limited network access fees (because it does not rely on the local telephone company for call origination), supports its low-cost business model. It took root among overshot customers that didn't need all of the features and the comprehensive customer support provided by the local RBOC. As of March 2004, more than one hundred thousand customers in the United States had signed up for Vonage's offering. Other players locked out of the local telephony markets, such as AT&T, announced plans to develop similar offerings (as did many cable companies, as we discuss later in the chapter).

What does the future hold for Vonage and other specialist providers seeking to use VoIP to drive growth in the residential market? Remember, successful low-end disruption requires having a shield of asymmetric motivation, which leads to incumbents fleeing from disruptive incursions at the low-end of their markets. The local telephony market bears some similarities to the airline market (discussed in chapter 6). Operators have huge, largely depreciated, fixed assets, which they try to spread over as many users as possible. Volume matters a great deal to them. This need for volume motivates them to respond to disruptive incursions. For example, Qwest began offering a VoIP product in Minnesota in December 2003. Verizon also announced plans to launch a competitive VoIP product. The fight for the low end of the market will most certainly hurt profitability. However, because incumbents are motivated to respond, the theories of innovation predict that VoIP providers on their own will not lead to a radical transformation of industry structure.

The ability of Vonage to easily enter the market illustrates another category of threats facing incumbents: the rise of specialist providers that can pick off pieces of an incumbent's business. IP enables the separation of the traditional interdependency between transport and services, allowing specialist companies to displace incumbents and capture a slice of industry value added. Overshooting creates the motivation for specialists to enter. IP creates the *possibility*.

The Threat of Decoupling

The vertically integrated, centrally coordinated, voice-based business model is poised to disintegrate. The Internet was one of the first steps in the separation of the service layer from the physical layer of the networks. Connection-oriented technologies were no longer necessary for high-quality secure connections. Centralized intelligence began to move toward the network's edge and became distributed and dynamic. Within the public Internet, the route that data takes to reach its destination is never predetermined. Technologies, such as virtual private networks (VPNs), create a delicate balance between connection-oriented and connectionless technologies. They establish flexible protected virtual network pathways that increase information security, but do not require deployment of costly dedicated networks.

The IP standard by definition allows interfaces to be specifiable, verifiable, and predictable. Further penetration of IP will enable a separation of the traditional, formerly necessary relationship between the transport provider and the service provider.[12] We call this *decoupling*. What would it mean?

The VCE theory suggests that the historical integration of transport and services must come at a cost. Remember, integration helps provide high functionality and reliability, but it limits the ability to provide customization and convenience. In a decoupled world, users could purchase customized and configurable services from specialized service providers instead of purchasing one-size-fits-all products offered by behemoths. Just as Vonage has emerged as a specialist service provider in the residential market for basic telephony services, a host of specialist providers could emerge offering advanced services such as call center management, toll-free services, and network management to small and large enterprises.

Decoupling would create opportunities for new horizontal business models across customers and markets to emerge where incumbents previously erected impenetrable barriers. These horizontal business models would offer products and services that piggyback on the connectivity and transport capabilities of the underlying physical infrastructure. The locus of integration would shift from the interaction between transport and services to critical subsystems and the interfaces that drive convenience and customization.

A widespread displacement could leave incumbents looking like hollowed-out shells of their former selves, relegated to providing "naked transport." They would bear a stark resemblance to their cousins in the electricity industry. In this scenario, specialists could build asymmetric skills. The lack of full integration that was a historic disadvantage would flip to becoming an advantage.

What could stop this evolutionary path? The government could somehow stop specialists from entering. Incumbents could develop the requisite skills to thrive in a decoupled world. But incumbents face a tough challenge. Existing strengths would be weaknesses. Internal response would seem difficult. Incumbents might have no choice but to set up separate organizations to develop the requisite skills.

VoIP Summary

Overall, the growth of wired data networks and VoIP presents both threats and opportunities to incumbent service providers. VoIP could be used in three different ways. Incumbents will naturally use it as a sustaining innovation to lower costs and add new high-value services. Entrants will continue their low-end attack, but incumbents are likely to fight back. And forward-thinking companies—either incumbents or entrants—could use VoIP as the basis of a new-market approach that makes it easier for nonconsuming companies to run world-class phone systems.

In short, it would seem to be just a matter of time before voice and voice services migrate entirely to a data context. Although incumbents seemingly have the skills and motivation to make that transition and play in this emerging data world, the transition would not be without

INDUSTRY SUMMARY

VoIP

Signals of Change: The ability of individuals and network managers to customize telephone services and the entrance of specialists, such as Vonage, that offer low-cost solutions indicate disruptive developments.

Competitive Battles: Incumbents are motivated and capable of adopting IP technologies and responding to incursion. Specialist service providers could develop different skills.

Strategic Choices: Choices to watch are entrants finding ways to create their own networks, incumbents setting up separate organizations to develop specialist service, and incumbents "cramming" IP.

---

risk. Response tactics are likely to depress profitability. Even worse, IP technologies will increasingly enable the separation of transport and services, creating opportunities for specialist providers to take sizeable pieces of incumbent business.

Incumbent telecommunication providers need to worry about more than specialist providers such as Vonage. Incumbents are being challenged from multiple fronts. The next section shows how blurring lines between cable companies and phone companies could put further pressure on incumbents.

2. Cable Companies: How to Create Asymmetries

Plagued by a less-than-stellar customer service reputation and constantly under inquiry for dubious pricing tactics, cable companies have earned a special place in the hearts and minds of many Americans. However, the cable companies that were once the punch line to endless late-night jokes have been reborn and thrust onto the front lines of the battle for local residential telephony and broadband data services.

Provisions of the 1996 Telecommunications Reform Act (discussed in chapter 4) granted cable operators permission to enter new markets previously off limits to any company other than an RBOC. Cable companies, redefined as multiple service operators (MSOs), could now upgrade their existing infrastructure to offer voice and data services.

The section explains why MSOs have increased their chances of creating new growth by moving from circuit-switched solutions to IP-based solutions. However, their continued focus on the largest market could be a flawed strategy, because it forces incumbents to fight. Picking an exact winner in this head-to-head battle is unclear. The battle will be bruising. The tactics combatants use during battle could result in the winner's rewards being a spoiled business with eroded profits. A more disruptive path, which could still emerge, would take advantage of asymmetries of motivation to increase the ultimate odds of winning an attractive prize.

Challenges in Technology and Market Selection

When an MSO decides to offer local telephony services, it has to choose its target market and its delivery mechanism. The natural target market, and the one chosen by every MSO to date, is the large and lucrative primary-line telephony market. Given the existing state of IP-based technology and the belief that PSTN-level quality was what customers required, early movers used circuit-switched solutions. However, the VCE theory explains why most companies have sensibly migrated toward IP solutions.

The two delivery technologies have important differences. Circuit-switched technology is more expensive but more reliable. To use circuit-switched technology, a MSO invests in switching equipment at the cable head end and at the customer's premise. It sends a technician to the customer's premise to install specialized equipment. In essence, an MSO builds infrastructure that mimics the phone company's infrastructure, allowing it to provide service that mimics the quality of the phone company's service. The biggest difference is that the signals ride on the cable company's network instead of the phone company's wires. Pure IP-based solutions are significantly less expensive but are also prone to performance issues. A cable company sends voice as a data packet over a customer's high-speed cable connection to the Internet. The customer installs a gateway (similar to Vonage's device) in his or her home that digitizes the analog voice signals from his or her phone and encapsulates them in packets for delivery to their destinations over a mix of public and private data networks.

Cable companies that sought to enter the telephony market early naturally looked to circuit-switched solutions. As of 2003, IP technologies were not good enough to replicate the performance of basic telephony service from the phone company. There were reliability concerns. IP-delivered service could not match the PSTN's five 9s reliability and real-time functionality. It lacked line-powering capabilities. MSOs made a natural decision to use the best technology to target the largest, most lucrative market. But could this path—using circuit-switched solutions to target the primary line market—be the underpinning for a disruptive strategy? The answer seems to be no for two reasons.

First, targeting the primary line market pits the MSOs in a direct competitive battle with the local phone company for market share. The products are similar. Will the phone company choose flight when the cable company begins making inroads? Absolutely not. The phone company has a lot to lose, so it will be very motivated to fight back.

Second, using circuit-switched technology does not give MSOs any business model advantages in a direct head-to-head battle. Why? Circuit-switched technology does not offer an MSO much of a cost advantage. Thus the preferred MSO tactic is a bundling strategy, enticing consumers with the convenience of a single bill and lower prices from combining services they typically bought separately. The VCE theory can help to assess the value of this bundling strategy by answering two questions:

1. Will the bundle truly cost less to produce? Through the lenses of the VCE theory, bundling adds value when stitching the bundle together requires unique understanding of interactions between bundled components. In other words, if there are interdependencies in production, an integrated firm can create an offering that a collection of focused specialists could not match. If there are no interdependencies, a bundled service is not inherently better or cheaper than two separate services.
2. Will consumers uniquely value the bundle? The theory suggests consumers value bundles when they consume elements of a bundled offering either simultaneously or in an interrelated manner. If this interdependence of consumption does not exist, customers have no natural reason to purchase the bundle.

From this perspective, there would seem to be limited benefits in bundling circuit-switched telephony and cable television. Although both services run over the same wires, they require completely different sets of equipment. The VCE model, therefore, would characterize consumer savings on an MSO's bundle as pure marketing promotions, not true business model advantages. The only additional savings to the customer is a first-class stamp each month, and even that savings disappears with online bill payment. Because both the MSOs and the RBOCs have similar models—heavy up-front capital investment with high gross margins over large customer groups to amortize those investments—MSOs can't use this approach to take advantage of business model asymmetries to make money at price points that telephony operators would find untenable.[13]

In short, using circuit-switched technology as a primary line replacement does not seem to create or harness asymmetries. A bundling strategy would only be a source of competitive advantage as long as a company was willing to pump promotional marketing funds toward subsidizing the bundle.

By early 2004, most cable companies seemingly came to this realization and began to migrate to using IP-based technologies. Time Warner launched a trial VoIP product in Portland, Maine, and Raleigh, North Carolina. Comcast planned rollouts in several big cities. Cox, one of the pioneers of circuit-switched telephony, decided to use IP in new markets.

How does the IP route look through the lenses of the VCE theory? There does in fact seem to be interdependence in the provision of IP-based telephony over a cable wire. The additional cost of providing IP-based telephony is quite low because the MSO can use the same infrastructure it uses to deliver high-speed access to the Internet.

Because of this interdependency, this approach has more disruptive potential than the circuit-switched approach. There are still two problems, however. First, MSOs are attempting to stretch a still not-quite-good-enough technology to a highly demanding market application. As discussed in chapter 3, these efforts often falter, especially in the face of steady, incremental improvements in existing technologies. Second, this strategy will still provoke a competitive response by RBOCs. By early 2004, the RBOCs had indicated two response tactics. First, as noted, they were developing their own VoIP offerings. Second, they were launching tit-for-tat strikes, partnering with direct-to-home satellite television providers to go after the cable company's home market.

Of course, there is one key asymmetry between the markets: It is technologically easier for a cable company to add a voice product than it is for a telephone company to add a video product (at least as of 2004). But note that the cable company's video product is a stand-alone offering in this model, appearing only as a line item on a bill. In fact, in many markets, MSOs still send out separate bills, one for TV and data services, the other for telephony services. Furthermore, cable companies seem to be more interested in the race for digital and on-demand services than in the telephone battle.

Cable companies striving to offer voice services are following a sustaining, not a disruptive, strategy. This statement does not imply that we think cable companies will fail. They could outspend their rivals. They could eke out incremental market-share gains. We are merely suggesting that this approach almost guarantees a head-to-head battle of similar services. These types of battles typically cost a lot of money, and the victor's "reward" is severely damaged profitability.

The next section shows the road not taken, illustrating how a different set of choices could lead to a very different outcome.

The Road Not Taken: IP as a Second-Line Replacement or as a Basis for New-Market Disruption

A company that establishes a disruptive foothold and marches up-market has a greater chance of actually reaching its destination than a company that tries to make the jump all at once. What path could MSOs choose that would begin the steady, incremental improvement process? Is there a way to use IP to establish a foothold that takes advantage of asymmetric motivation and builds asymmetric skills?

First, let's assess the potential for offering a good-enough product at lower prices to overshot customers whose defection would not incur incumbent response. Cable companies could, for example, use IP to offer cheap second lines to customers. IP is certainly good enough to offer a relatively cheap, fully functioned product (for example, telephony service that includes basic calling-related features such as voice mail, call waiting, caller ID, and so on) to customers.

This could be a potential low-end disruptive growth opportunity—a lower-cost business model targeting the incumbent's most overshot customers. Targeting the second line might take advantage of asymmetries of motivation. By 2004, the secondary-line market was already in free fall. Parents bought wireless phones for teenagers. High-speed Internet access eliminated the need for a dedicated line for modem connections. RBOCs would not happily give away more second lines, but they might not even notice MSOs making inroads because the market was already shrinking.[14]

What about a new-market growth opportunity? At first glance, this path appears difficult. Who is a nonconsumer of voice services in the United States? Just about everyone has a telephone. However, there could be opportunities for industry players to build relatively simple voice, video, and data packages that allow people to do things that historically were impossible. For example, an MSO could offer a package targeting baseball aficionados, bundling together Major League Baseball's MLB.TV streaming video offering with a VoIP-based phone service that would allow friends around the world to chat while they all watched the game (and even see each other with the help of a simple camera). Although the video and voice offerings on their own would seem technologically limited, fans might welcome such offerings because it would allow them to watch games in new contexts and interact with like-minded fans in ways that were historically impossible. Bringing voice services into new contexts that increase people's ability to interact and be entertained could be the basis for new-market growth. There certainly could be other ideas that have more growth potential than this one. The key is that the idea must make sense when viewed from a jobs-to-be-done perspective but make less sense when viewed from the incumbent's perspective (either because of the small initial market size or the unique skills required to make the idea a reality).

A cable company could use either approach to learn about the telephone service market while increasing the slope of the technology's improvement trajectory. Improving technology could eventually allow customers to use IP for primary line service. MSOs could add new converged services, such as unified messaging (the combination of voice, e-mail, and fax). This would allow cable companies to build the skills necessary to win a full-on attack. Following this approach would make RBOC response more difficult, because MSOs would have created new business models that profitably serve less demanding customers.

Cable Telephony Summary

Figure 10-1 summarizes the choices facing MSOs. This depiction is typical. Firms can try to enter an existing market with a competing sustain-

FIGURE 10-1

Options for MSOs

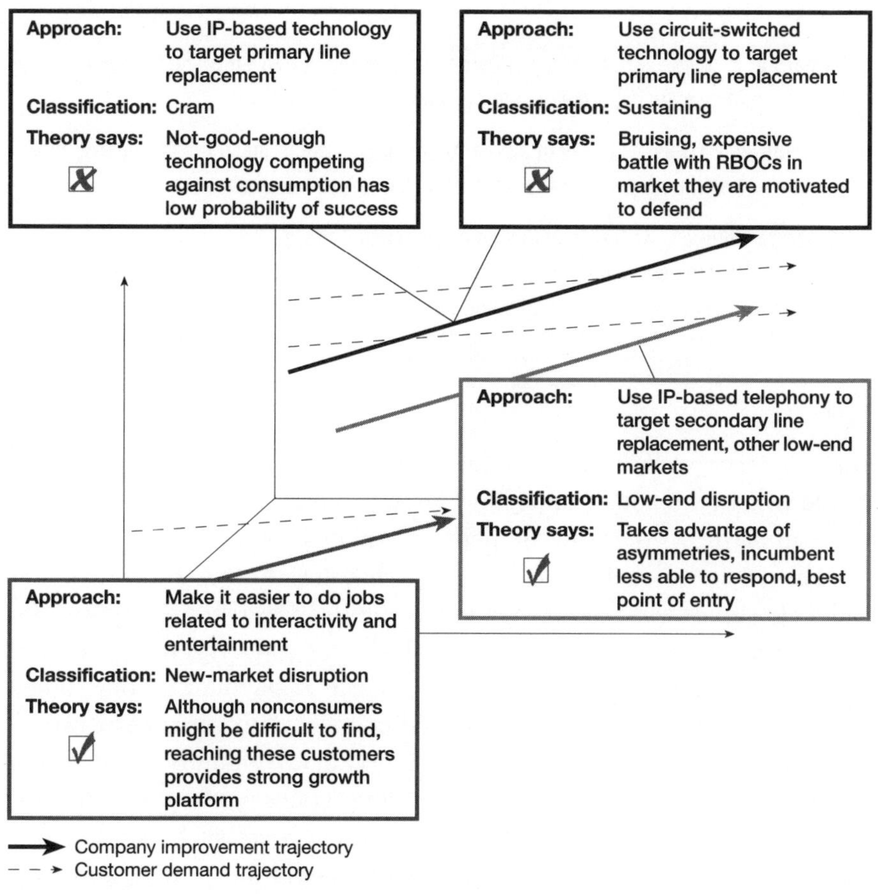

ing innovation. If they have an innovation that has limitations but offers a new benefit, they can cram it into a context where it still has some limitations or use it disruptively.

Theory always advises betting on disruption because growth takes advantage of asymmetric motivation, giving the attackers time to develop unique skills. This section showed that crafting a viable, disruptive strategy is probably possible—but more than in other industries, natural economic and technological forces in telecommunications services make the disruptive route a hard one to take. It is hard to see, if

INDUSTRY SUMMARY

Cable Telephony

Signals of Change: Change in regulatory context facilitated entry. Second-line customers could be overshot. Nonconsuming contexts exist.

Competitive Battles: A battle of similar technologies would mean no asymmetric motivation unless interdependence in production led to lower delivery costs.

Strategic Choices: Entrants (cable companies) could target low-quality second lines to overshot customers or unique entertainment-based bundles to nonconsumers, facilitating their creation of unique skills and business models.

the competitors clash in a sustaining innovation fight, who will win the battle when both combatants have good products and deep pockets. All you can predict is that the competition will be brutal.

The trick is locating a company that has found a context that uniquely values what the company has to offer. Overshot telephony customers might welcome a cheap, albeit limited, product. A unique bundle that made it easier to get done key jobs (such as having fun) or brought communication to new contexts could connect with customers, even if each piece of the bundle had some technological limitations. If MSOs were able to deploy IP-based telephony service with a significantly lower-cost business model, integrate it with services such as video in a way that the phone company (as of 2004) could not match, and were not subjected to additional regulatory intervention, a theory-based assessment would have to tip in favor of the attacker.

Cable companies are natural competitors because their wires run into most American homes. Transmitting data increasingly involves more than wires. The next section evaluates developments in wireless data.

3. Wireless Data: Flawed Initial Strategies Could Turn to Co-option

After discussing VoIP and cable, which largely rely on wired networks, a natural place to turn is wireless data. This section shows how despite flawed initial deployment strategies—such as massive investments that attempted to cram not-good-enough technology to the most demanding applications—incumbents seem likely to co-opt next-generation wireless technologies such as 802.11 and use them to augment existing business models. But there's potential peril for the incumbents as 802.11 and VoIP converge and new technologies such as 802.16 emerge.

3G and NTT DoCoMo: Cramming and "Oh, Hello Kitty"

During the 1990s, wireless voice and the Internet both grew exponentially. Not surprisingly, people grew increasingly excited about combining the two—accessing data using wireless technologies. Existing wireless companies saw the opportunity. But they framed the challenge as, "How do we stretch the technology to meet the data needs of deep-pocketed businesspeople?" This set most companies on a development path involving billions in investment to disappoint customers. It was a classic cram.

In the late 1980s, engineers began developing so-called third-generation (3G) technologies as a sustaining innovation to increase wireless service providers' utilization of their scarce spectrum allocations for voice services.[15] Additionally, 3G offered higher transmission speeds, making it a potentially useful means to send data applications. However, because engineers designed 3G primarily to improve voice services, its data capabilities were somewhat limited.

Executives had a choice. They could introduce simple, low-bandwidth applications, proceed with the voice upgrade, and incrementally introduce higher-performing applications. The specific applications customers would want were unknown. The markets appeared small. Or they could try to stretch 3G to provide high-end applications such as wireless videoconferences. Customers certainly would demand such applications. The markets appeared large. But the technology was not quite ready.

To most executives, the choice seemed clear. Developers would have to modify 3G to enable higher transmission speeds capable of delivering advanced data services that could reach the most lucrative markets. European operators spent billions of dollars acquiring scarce spectrum licenses in auctions designed by game theorists. Auction "winners" took on heavy debt loads that only furthered their desire to reach up-market. How else could they justify the huge up-front investments?

So operators spent billions more trying to improve the technologies. Developers tried to modify 3G to enable the seamless provision and delivery of up to two megabits per second of data—more than ten times faster than the speed it was designed to deliver. The tolerance for error was low. Nonfunctioning products and services cost demanding users—such as mobile salespeople—real time and money. Imagine the ire of a salesperson whose streaming video connection locked up when she was seconds away from closing a multimillion-dollar deal.

Initial excitement waned as it became clear that 3G technologies could not meet expectations. By 2002, observers decried the license investments as a huge waste of money. The demanding applications hadn't materialized—the technology just wasn't good enough. Many within the industry began to pin their hopes on fourth- and fifth-generation technologies to deliver the vision of a high-speed wireless world.

Most operators fell prey to the siren's call of cramming. Instead of embracing simple opportunities that would delight customers, they tried to cram a technology to meet the needs of the most demanding customers.

One operator, NTT DoCoMo in Japan, showed how a path focused on simple services—targeted at nonconsumers of wireline data services such as teenage girls—could drive disruptive growth. DoCoMo's i-Mode service began with simple applications such as novel ring tones and cartoon characters such as Hello Kitty that appeared on the phones' screens.[16] By giving customers something they could never have before, DoCoMo delighted customers with even the most rudimentary of applications.[17] DoCoMo didn't know exactly what applications its customers would value, so it created a business model that encouraged rampant experimentation among third-party content providers. Providers would further develop applications that connected with consumers. Providers would cut off applications that failed. As of September 2003, DoCoMo had close to 40 million delighted i-Mode customers enjoying rich graphical experiences.[18]

Western operators soon followed, capturing growth from simple services such as short messaging service (SMS), which allowed users to blast brief text messages to their friends. For example, in the United Kingdom, users—including anxious students exchanging A-level test results—sent more than 1.6 billion text messages in August 2003.[19]

WiFi: Disruptive Growth and Then . . . Co-option?

As DoCoMo demonstrated how the path to profitable growth in wireless data success began in simple applications, an entirely new way to transmit data wirelessly emerged from the computing and networking industries. Technologists call the technology by its Institute of Electrical and Electronics Engineers specification number, 802.11; nontechnologists call it WiFi (short for wireless fidelity). WiFi is a fundamentally different way to deliver data wirelessly. Technologies based on 3G are relatively slow but cover a relatively wide area. WiFi-based technologies are relatively fast but cover a relatively limited area (most WiFi transmitters have a functional range of about 300 feet).

One big advantage of WiFi-based products is that, unlike 3G, WiFi does not require obtaining scarce licenses from national governments. It operates in so-called unlicensed spectrum set aside by the U.S. government to encourage experimentation and innovation.[20] In 2003, the government's desire to foster competition appeared to be coming true. These high-speed, shorter-distance wireless technologies created a sort of Wild West of innovation in the wireless data-networking model. WiFi articles with headlines such as "The WiFi Revolution" were featured on the covers of leading business magazines.

Companies began using WiFi to set up wireless local area networks (WLANs). Users would install a simple card in their computers (or use a computer that had the functionality built in, such as in laptops powered by Intel's increasingly popular Centrino line of processors) and receive a signal from an access point connected to the wired Internet. Individuals set up WLANs in their homes, freeing them from the limitations of wires. WLAN providers began sprouting up to service heavily trafficked areas (known as "hot spots") such as airports, cafés, and downtown business districts.

This is classic new-market disruptive growth. WLANs bring network connectivity to previously unimaginable places. Access points are easy to set up and simple to use. They make it easier for people to do something they historically prioritized—accessing data anywhere, anytime, at very high speeds

Who exactly are these WLAN providers? Companies are experimenting with various approaches to build WLANs. Boingo, for example, launched by Internet service provider pioneer Sky Dayton, is attempting to build a business model aggregating microcarriers ranging from a wireless access point in someone's living room to the network in a local hotel. Mesh Networks is attempting to build a network with no central architecture—devices can communicate with other devices. Countless companies with names like Flarion and Navini Networks are working to improve existing products and create myriad complementary products and services.

To move up-market, these providers need to overcome some serious technological challenges. Issues related to limited distance (such as the technology's range), presence management, and security all make crafting a profitable business model challenging. Wizened industry veterans suggest that it will be next to impossible to overcome these hurdles. Regulation and physics, they note, constrain WiFi's improvement trajectory. Perhaps. But history suggests that when there is ample motivation—as there certainly is, in this case—innovators will eventually overcome these barriers. The *X*-factor could be the government, which conceivably could impose barriers that limit the amount of available spectrum for these novel solutions.

Assuming, however, that the technology continues to improve, would this be good new or bad news for incumbent service providers? Would this be a circumstance in which co-option is a possibility? The answer ultimately will depend on choices made by innovators as they coalesce around a business model and further develop WiFi solutions. As of early 2004, evidence overwhelmingly pointed to a repeat of the wireless voice story—a disruptive application co-opted by incumbents. However, the convergence of 802.11 and VoIP and the emergence of next-generation technologies could change this assessment.

The Choices That Could Power Near-Term Disruption

Early market developments for WiFi relied on the shield of asymmetric motivation—the markets were ill defined and too small to matter to multibillion-dollar companies that were focused on finding a way to make 3G work. To cause the incumbents to ignore WiFi's growth for a long time, entrants would need to develop unique business models and skills. WiFi operates in a largely freestanding value network. Of course, a wireless data provider must connect to a wired network at some point. But this interaction is at a true point of modularity. This modularity gives entrants the ability to develop business models not based around minutes of use. They could offer much lower prices because they would not have to build massive networks.

A new crop of competitive access providers (CAPs) could cobble together enough wireless access points to create some semblance of national coverage. This could pull data traffic into a new, wireless context.

Next, WLAN providers could develop a way to offer voice services (a sort of a wireless VoIP, discussed later in the chapter), cementing their disruptive threat to incumbent providers. This disruptive path would ultimately leave incumbent network providers with an important backbone role. But they would largely lose control of the customer and would be significantly less important industry players.

Motivation to Fight Early, Complementary Strengths Likely to Power Sustaining Scenario

Would incumbents ignore WLAN developments long enough for entrants to get so far down this path that response will come too late? In 2003, signs began appearing that incumbents were not afflicted with asymmetric motivation. A number of carriers such as T-Mobile, Verizon, and AT&T began experimenting with novel ways to use WiFi to augment their existing networks. T-Mobile started to sell WiFi access at Starbucks coffee shops. Verizon said it would allow existing DSL customers to utilize WiFi access points it planned to install in its pay phones in some cities. AT&T, IBM, and Intel formed a separate company called Cometa Networks to wholesale WiFi services (although Cometa shut down in May 2004 after failing to raise enough capital for expansion).

Furthermore, it is unclear if competing in localized areas really requires distinct skills from overseeing a wide-branching network. Successful WLAN providers need to solve problems such as bridging between networks, managing numerous individual customer accounts, and processing small transactions. Large incumbent operators must overcome similar types of problems, indicating that their skills might match this opportunity quite well.

If entrants cannot create an asymmetric situation, the incumbents will have a good chance of co-opting WiFi. What would a co-opted world look like? The CAPs emerging in hot spots would provide a complementary high-speed data network able to handle local bandwidth-intensive traffic (similar to the way in which regional jets augment existing hub-and-spoke networks). Incumbent carriers would keep the technology at arm's length until it became good enough to deliver high quality and reliable service within their networks. Eventually, these CAPs would prove attractive acquisition targets to existing operators looking to augment their coverage. With superior resources and network management knowledge, incumbent providers ultimately could capitalize on the growth. Emerging players that explicitly seek to develop complementary services to make themselves look highly attractive to potential suitors would find the most success in this scenario.

The Future Disruptive Path

Even though incumbents seem poised to co-opt 802.11 technologies, two future developments could set the stage for true disruption. First, the convergence of VoIP with wireless technologies could allow providers to develop truly novel business models in which users pay a one-time charge to receive unlimited voice calls and Internet access. For example, Dartmouth College has blanketed its campus with 802.11 access points. In September 2003, it provided "softphones" to all of its first-year students. Students could plug the softphones into their laptops and make and receive telephone calls for free anywhere on campus.

Incumbents do have some strengths to utilize here as well. They could come up with a truly integrated solution that offers customers one phone number that works wherever they are, flipping between wireless and wired networks depending on the customer's location. But the new business models bound to flourish have serious organizational implications, likely requiring a separate organization for commercialization. Also, incumbents lack some of the skills that are required to thrive in this world, including skills related to software development. The sooner the convergence between 802.11 and VoIP happens, in other words, the more difficult co-option will be.

The second development to watch is the development of new wireless technologies, such as 802.16, also known as WiMax. Early tests show that WiMax products will provide access speeds of up to seventy-five megabits per second at a functional range of thirty miles. The technology is still in the development stages, with early iterations suffering from limitations related to mobility and roaming. However, if development continues as projected, this could be a real game changer. New providers could emerge to offer truly converged services at very low prices. Cobbling together a national network is a much easier task given 802.16's technological characteristics. One high-potential path an entrant bent on disruption could follow would be to target initial markets that are either too small to matter to incumbents or in geographies that incumbents do not serve. For example, developing a low-cost WiMax network in underserved areas in developing countries would not lead to immediate response. It would allow an entrant to hone a business model and supporting processes that could be powerful competitive weapons. One possible outcome is the emergence of WiMax or other next-generation technologies as alternative ways for providers to offer basic connectivity. This would put further pressure on the provision of transport and hasten the value shift toward the provision of decoupled services, discussed earlier in this chapter. One open question is whether the government (specifically the FCC) will consider 802.16 in the same way it has the other 802.*x* technologies.

Wireless Data Summary

While we write this book, the future of wireless data remains unclear. Current evidence points to the sustaining scenario, in which incumbent service providers would continue to be motivated to co-opt WiFi technologies. However, a disruptive scenario remains in which a class of entrants could break free, set up a freestanding value network, develop a non co-optable business model, create asymmetric skills, and disrupt incumbent wireless and wireline service providers. Companies to watch closely are ones like Mesh Networks, which seems to be explicitly following this strategy. Mesh might not get it right, but a company that follows this path could stumble upon a legitimately disruptive strategy.

INDUSTRY SUMMARY

Wireless Data

Signals of Change: WiFi networks emerged in new, nonconsuming contexts. To date, the government has not constrained ability.

Competitive Battles: Early providers took advantage of incumbent disinterest in small markets. The key question is whether entrants develop unique skills and business models. Context makes flight unlikely.

Strategic Choices: Entrants need to maximize independence of value networks and develop new ways to reach new customers. Finding ways to combine 802.11 and VoIP is critical.

The chances for disruption go up as VoIP and wireless technologies converge and as next-generation wireless technologies continue to develop.

Readers can use our tools to evaluate the decisions firms make as they march forward. They can watch to see the business models entrants choose and the skills entrants develop and observe how incumbents choose to fight. Pay careful attention to emerging companies that follow the right preparation regimen and set up freestanding value networks, even if they start in nonobvious markets. Also watch to see if entrants are building to disrupt or building to sell to incumbents. As always, theory can help provide deep insight into ultimate outcomes before corroborating data exists.

4. The Fringe: Competitors from Strange Places

So far, we have looked at pressures on incumbents from wired and wireless data providers and from the other wire that comes into almost every U.S. household. All of these threats are somewhat obvious. And incumbents seem to have a good chance of fending off each of them. Predicting industry change requires unearthing the unexpected. It requires looking for companies bringing things into such a different context or doing things in such a different way that incumbent response would just not be possible.

This section touches on two such examples—instant messaging (IM) and Microsoft. These examples may seem random or trivial, but remember, what is trivial today can be world-changing tomorrow. For each development, one could chart a natural improvement trajectory that results in the emergence of very surprising competitors.

IM: Toy or Disruptive Platform?

Millions of teenagers worldwide spend hours each day communicating with each other via IM services. The free applications offered by AOL (AOL Instant Messenger), Yahoo! (Messenger), and Microsoft (Windows Messenger) have been quietly and quickly downloaded onto millions of computers around the globe. Once operational, users can type a message (often littered with acronyms such as LOL, TTYL, and AAMOF[21]), press enter, and almost instantly transmit the message to their friends' computers or portable devices.

This is disruptive growth. IM brings real-time communications into a new context.[22] People can communicate unobtrusively without the possibility of people nearby (such as, perhaps, parents) hearing a sensitive snippet of conversation. IM allows teenagers to gossip and chat to their hearts' content, without being overheard, while their parents think they are doing homework. In this context, not being able to use your voice is a benefit, not a burden.

IM is relentlessly improving. Colleagues halfway around the world can jointly create and edit a PowerPoint presentation. People can use IM on mobile devices. Users can press the talk button and broadcast their voices across the Internet to their friends. Teenagers can purchase a cheap Web camera, connect it to their PCs, press the Web cam button, and voilà—a simple videoconference. IM also has presence management capability, meaning that a user's friends know when she is on- or offline.

None of these applications are very good. Voice transmissions echo and sound tinny. The image quality of a videoconference is poor. Telecommunications service provider executives might scoff at the relatively low quality of these services, but they work and they are free. Users are delighted. Executives ought to pay attention.

IM's growth and improvement is worth watching because companies that play in the IM market could develop business models that just don't make sense to telecommunications companies. Companies offering IM do not look to directly monetize the usage of IM; communications itself is not their raison d'être. Their goal is to use IM to increase the relevance and usage of their core products and services. Companies are solving problems and creating skills that bear little resemblance to incumbent service provider skills.

This could be the ultimate decoupled service, teeming with asymmetries. Whether or not IM really heralds changes to the telecommunications industry remains to be seen. But watching its up-market march will allow us to see if it portends peril for existing providers.

Another Potential Low-End Disruptive Business Model: Microsoft

How can a book about innovation and technology not have at least a small section dedicated to Microsoft? Microsoft is playing in the IM market. But that is not all Microsoft is doing. The latest version of its operating system, Windows XP, is compatible with the session initiation

INDUSTRY SUMMARY

Fringe

Signals of Change: IM has created new-market disruptive growth by bringing communications to new contexts; Microsoft could develop a simple, low-cost solution.

Competitive Battles: Both developments involve fundamentally different business models that will make incumbent response difficult. Providers are building skills not shared by incumbent providers.

Strategic Choices: Big questions are: Will incumbents continue to ignore such developments? What path will companies providing these services use to move up-market? What business model will they use? Will they use the right preparation regimen?

---

protocol (SIP). SIP is a relatively simple standard for VoIP transmissions that works on most communications networks.[23]

What does this mean? Microsoft's embedded functionality could make it easier for users to use their computers as telephones. As Microsoft itself noted on its Web site, "every copy of Windows XP that ships is essentially a telephone handset."[24] In essence, companies could create applications that mimic the functionality of an IP phone in a computer environment. Does this have a disruptive ring to it?

Because Microsoft understands better than any company in the world how to create a well-functioning operating system, it conceivably could integrate voice technologies better than anyone else. This interdependence of production could allow Microsoft to offer a lower-cost telephony solution compared to incumbent service providers. This could be a legitimate low-end disruption. Incumbents would lack the requisite skills or knowledge to respond effectively.

As of March 2004, Microsoft disavowed any interest in becoming a telecommunications player. It announced it was merely interested in new service-based revenue streams and supporting other service providers. It quietly introduced SIP into XP. It was waiting to see how consumers used the functionality before deciding its next move. Nonetheless, analysts looking for the next big disruption must carefully watch to see if Microsoft quietly launches a disruptive attack.

After the wreckage of the past few years, the easy answer would be to always bet on incumbents. Incumbents have obvious strengths. They own the networks. They know how to work with regulatory agencies. They have the resources.

Indeed, the telecommunications industry has historically seen waves of disruptive innovation that in other contexts might have systematically changed the structure of the industry. Specialist firms have entered the market. Disruption has flourished at the edge of the network (for example, equipment) and in adjacent markets (for example, enterprise data markets). However, disruption has not shaken the core of the telecommunications industry as it has industries such as retailing, computing, and electronics.[25] *The basic business model of tight vertical integration and centralized management, control, and coordination has largely remained unchanged and unchallenged. Nothing has fundamentally reshaped the industry since the introduction of the telephone.*

But, as Bob Dylan said, "The times they are a-changin'." Even if incumbents co-opt each development discussed in this chapter, they face competitive threats on many sides. Independent wireless companies, cable companies, and specialists riding the wave of decoupling can all nibble at incumbent markets. Death by a thousand cuts is a real possibility. And that would be a relatively good *outcome. The worst scenario from an incumbent perspective involves a sudden shift from centralized-wired-voice to decentralized-wireless-data, with incumbents ruling an old world that becomes increasingly irrelevant.*

Therefore, incumbents to bet on are those that embrace the forces early and seek to utilize technology to create disruptive growth businesses. Companies that wait and see often cease to be. Entrants and nontraditional industry players to bet on are those that make the right set of decisions, such as avoiding the trap of large, obvious markets, developing asymmetric business models, and creating freestanding value networks where they can build asymmetric skills.

There is of course a caveat to any forward-looking scenario. Because the government still plays a large role in the industry, its actions could heavily influence the course of industry change. For example, a government clampdown on the use of unlicensed spectrum would make WiFi look a great deal less promising. Similarly, decisions could seriously affect specialist VoIP providers and cable operators.

The telecommunications chapter has five general lessons:

- *The analytical approach laid out in chapters 1 through 3 can bring order to innovations in complicated industries.*
- *One of the most valuable uses of theory is to focus attention on the developments that bear watching, giving futurists the chance to correctly interpret signs that signal an innovation's and an industry's evolution.*
- *Be wary of overestimating the chances that an innovation will overthrow incumbent leaders. Look at the co-optability of the innovation and the motivation of incumbents to fight the disruption rather than to flee it.*
- *Highly interdependent networked industries provide avenues for co-option.*
- *Developments in the fringe always bear watching.*

Notes

1. Our analysis is clearly not comprehensive. Other important technological developments include optical technologies, softswitches (a software-based way to switch voice calls from one network to another), metro-area Ethernet providers, electrical companies providing Internet services, and so on. Also, our focus on service providers means we give short shrift to equipment providers. But providing a comprehensive look into the future of every aspect of the telecommunications industry would literally require an entire book.

2. This data comes from various sources including company annual reports and quarterly filings, OneSource, and various analyst reports. Particularly useful analyst reports included: Viktor Shvets and Andrew Kieley, "RBOCs: Initiating Coverage: '. . . But He's Got My Switch!'" Deutsche Bank Securities Inc., 22 November 2002; Simon Flannery, Jeannette Baez, Paul Enright, and Mark McKeown, "Wireline Telecom Services: Trend Tracker: Rebound? Not Yet in Telecom," Morgan Stanley, 21 November 2002. Reports accessed from Investext. Wireless revenues do not include AT&T Wireless, which AT&T spun out in 2002. It does include Cingular (through SBC and BellSouth's equity ownership), Verizon Wireless, and Sprint's PCS service.

3. Of course, many of the wireless calls users view as having low marginal costs actually have quite high average costs once service fees are amortized over the minutes of actual use.

4. This of course would not be true for *all* market tiers. There will always be very demanding users who will not tolerate a dropped call, such as salespeople or companies that rely on their telecommunications systems for mission-critical applications.

5. Our thanks to Michael Putz, Donna Soave, and Kevin Kennedy of Cisco Systems for sharing this story with us. Interestingly, whereas Cisco Systems marketed these technologies, many of the firm's capabilities were acquired during an aggressive acquisition spree in the 1990s. The acquisition of many companies with innovative technologies fueled Cisco's rapid growth.

6. There are two basic types of networks: circuit-switched networks and packet-switched networks. Circuit switching essentially involves creating a direct electrical connection between two points. Circuit-switched technology is highly reliable but relatively inflexible. In a packet-switched network, there is no direct end-to-end connection. Data gets broken down into small packets that are blasted across the network, only to be reassembled at the final destination. Historically, packet-switching technologies were less reliable but much more flexible. For the purposes of this chapter, we make the simplifying assumption that packet-switched networks primarily transmit data and circuit-switched networks primarily transmit voice.

7. We draw a distinction between delivering packets over any-to-any systems and delivering packets over connection-oriented systems. Frame relay, asynchronous transfer mode, and multiprotocol label switching all allow companies to send packets over point-to-point technologies. This grants them greater reliability, but obviates the flexibility and low-cost benefits of an any-to-any solution.

8. Lost packets lead to two problems: latency and jitter. Latency refers to the delay equal to the time it takes for a bit of data to travel from its origin through the network to its destination. Jitter generally refers to the variation in quality of the voice signal resulting from the out-of-sequence arrival of a stream of packets. Public networks with no or low quality of service (QoS) do not guarantee the arrival of packets in the order they were sent, which can create distortions in voice transmissions. Network congestion and complex routing and translation challenges exacerbate latency and jitter.

9. VoIP providers would send basic voice traffic via Internet service provider (ISP) gateways to public networks. They would thus avoid costly fees telecommunications companies charge for interconnection because there was no way to segment out data to see where calls originated. These tariffs were traditionally very high for international calls.

10. Integrated companies have advantages when necessary functionality requires interdependence across the entire system. These interdependencies may only appear after the assembly of an entire system. Consider the development of a large software program. While many software companies attempt to develop their programs in modular fashion, with different sets of programmers working on different pieces of the program, the program cannot be debugged until all of the modules are combined. The combination of the modules into an integrated product often leads to unforeseen interactions and complications that only become apparent when the modules interact. Hence, a single enterprise is more likely to

develop the "black art" required to bring this new technology to market. This is a critical reason why integrated providers historically have had advantages in the telecommunications industry. You had to oversee a large-scale network to solve problems related to complex, tightly coupled, close to real-time interaction of multiple distributed systems interconnected across wide area interfaces. Only an integrated company could master the distributed systems-level interactions of the network. We would like to thank Michael Putz from Cisco Systems for helping us understand these concepts.

11. For example, customers are demanding that their service providers have the capability to differentiate between different types of data (such as real-time, mission-critical voice versus best-effort e-mail). To combat latency and jitter in a voice call, for example, VoIP traffic must be prioritized over other, less time-sensitive types of data. These class and quality-of-service distinctions have resulted in an unanticipated, but growing relationship between IP-based networking and traditional connection-oriented technologies. It is important to point out that this presents some constraints as well: Incumbents have to make sure that services leverage their existing asset base and technology platforms.

12. At a very high level, conservation of integration helps to explain this transformation. Previously, transport and service provision were interdependent. Using IP relies on routers. Because routers are not good enough, they have to be interdependent and integrated. But this means that the things that the router plugs into have to be modular and configurable. Hence the relationship between transport and services becomes torn.

13. A similar battle is emerging in the banking market between investment and commercial banks. Commercial banks have historically focused on lower-margin consumer and commercial loans. Investment banks have historically focused on higher-margin equity offerings, merger and acquisition activities, and advisory services. Leading commercial banks have begun to move up-market, taking increasing share of the higher-margin business offerings. The share of equity underwriting (when banks help companies sell shares of stocks) of America's three largest commercial banks—Citigroup, JP Morgan Chase, and Bank of America—grew from 12 percent in 2000 to 18 percent in 2003. Although there are no apparent interdependencies, commercial banks still have an advantage. Their business model allows them to make money on low-margin businesses. Their profitability increases even if they make less money than an investment bank on an individual high-end transaction. The world looks different through the eyes of the investment banks. Going into the commercial banking market looks structurally unattractive, given investment banking's cost structure and business model. So it is not commercial banking's bundle that allows it to gain share. Its business model makes it attractive for commercial banks to offer a price point that would look unattractive to the incumbents. See Jonathon Sapsford, "Banks Give Wall Street a Run for Its Money," *Wall Street Journal*, 5 January 2004.

14. Aggressively pricing a secondary line as a cheap add-on to data service would present some challenges. Since dial-up Internet access is one of the primary drivers of traditional second-line purchasers, cable companies packaging the sec-

ond line with broadband technologies (which obviate the traditional need for a second line) will have to creatively target high-speed consumers to drive adoption. Bundling voice as part of a data offering and positioning it as a relatively inexpensive second line will require creative marketing tactics to convert existing multiple line owners or create new ones. While recently second lines have been used for faxes or dial-up modems, perhaps an inexpensive line might expose a new and untapped additive opportunity.

15. For the purpose of simplifying the discussion, we use 3G as an umbrella term to represent the many different varieties of third-generation wireless technologies.

16. We found professor Jeffrey Funk from Kobe University in Japan to be a particularly knowledgeable source about i-Mode's history. See, for example, Jeffrey Funk, *Mobile Disruption: The Technologies and Applications That Are Driving the Mobile Internet* (Hoboken, NJ: Wiley, 2004). More of Professor Funk's astute work is available at <http://www.rieb.kobe-u.ac.jp/~funk/> (accessed on 1 April 2004).

17. Some academics discount i-Mode's success and argue that its commercialization path was largely a historical accident resulting from the unique nature of the Japanese market. Low activation charges and no roaming fees, they argue, inexorably lead companies to try to meet the needs of the mass consumer market, while European and American firms gravitate toward the business markets. Unlike the United States, where traveling business customers are highly coveted, the consumer market is equally important and the source of tremendous innovation. Also, the Japanese population is highly concentrated, making it less complex to provide reliable coverage. The smaller geographic footprint of this comparatively dense population has ruled out the need for roaming agreements and, in the view of the providers, elevated the appeal of the mass consumer market.

18. Subscriber numbers from Robert Budden, "I-Mode Getting Through," *Financial Times*, 16 September 2003. Interestingly, DoCoMo ultimately fell prey to the same forces that snared Western wireless executives. It tried to accelerate its jump to 3G services by developing a means to deliver streaming video called freedom of mobile multimedia access (FOMA). Few customers signed up. Taking advantage of DoCoMo's distraction, rival operator J-Phone introduced a relatively simple service that allowed customers to take snapshots and send them wirelessly to their friends. The application could run on J-Phone's existing network. Hundreds of thousands of users signed up. Once again, a simple application that delighted customers succeeded, whereas a great leap forward failed. See Jeanette Borzo, "Let Me Entertain You—With Mobile Phones, Users Can Get Their Kicks Wherever They Go—Tuning In to Interactive Games, Movie Clips and Virtual Pets," *Wall Street Journal Europe*, 8 March 2002; "Japan's DoCoMo Attempts to Catch Competitors with Camera-Phone," *Agence France-Presse*, 28 April 2002.

19. That works out to an average of more than 50 million text messages a day. Figures from the Mobile Data Association (MDA). See "59 Million Text Messages Sent Per Day During November," <http://www.mda-mobiledata.org/resource/hottopics/sms.asp> (accessed 6 March 2004).

20. That spectrum is in the 2.4- or 5-gigahertz range.

21. That's "laugh out loud," "talk to you later," and "as a matter of fact" for the uninitiated.

22. Another example of voice moving into a new context is interactive gaming. Once the lonely home of hackers and testosterone-filled teenagers, interactive gaming began to come of age over the last decade. Technology has improved dramatically, allowing companies to create artificial universes where thousands of users can participate simultaneously. Users flock to lower-tech sites where they can play relatively simple games such as Hearts. Next-generation video game systems such as Microsoft's Xbox and Sony's PlayStation 2 incorporate broadband capability, allowing consumers thousands of miles away to play popular titles such as *John Madden Football* against each other. And the people playing these games are not just teenagers—they are older and more affluent than ever before.

Like IM, gaming technology is moving up-market. The U.S. Defense Department has spent billions on "virtual" online training. Companies are beginning to realize that sophisticated simulations can be valuable training tools. Educational providers are recognizing that it might be easier to make fun educational than to make education fun.

As improvement continues, gaming companies could bring voice communications to a new context in which consumers again would welcome a relatively rudimentary product. Gaming companies provide communications in a very different way than traditional companies. Like IM providers, interactive gaming companies live in a largely freestanding value network where they only tangentially interact with existing service providers. They are solving difficult problems involving real-time applications with thousands of users. Improvements could move communications away from the classic voice paradigm. Voice would go from a core service to a useful tool to get other important jobs done. Perhaps these new providers could develop a way to make money that makes no sense to providers who are used to charging by the minute.

For more information, see Billy Pidgeon, Elif Akcayli, Jay Horwitz, Ed Kahn, Joseph Laszlo, and Andrew Peach, "Connected Consoles: Online Games Initiate New Distribution Channels," Jupiter Media Metrix, Volume 2, (2001); Michael Macedonia, "Games Soldiers Play," *IEEE Spectrum*, March 2002, 32.

23. George Malim, "Opportunity Knocks," *Communications International*, July 2002, 16.

24. Microsoft Corporation, "Microsoft Windows XP Real Time Communications Opportunities," <http://www.microsoft.com/serviceproviders/voiceservices/> (accessed 21 August 2002).

25. The words we choose are important here. As we have repeatedly noted, there have been important disruptive developments. Cisco's development of the router, ROLM's creation of the PBX market, and the creation of wireless technologies all meet the characteristics of disruptive developments. None of these innovations led to a widespread change in who provides telecommunications service on the order of the change that resulted from the introduction of the telephone.

CONCLUSION

WHAT'S NEXT?

To close *Seeing What's Next*, we would like to step back and recap where we've been, suggest areas for further investigation, and provide some parting thoughts.

Where We've Been: Using Theory to Analyze

Seeing What's Next detailed a way to use the theories of innovation to predict industry change. In chapters 1 through 3, we presented the core of our analytical process. The process involves three iterative steps:

1. Look for signals of change that point to changes to an industry's context or companies using new ways to reach nonconsumers, undershot customers, and overshot customers.
2. Evaluate competitive battles by comparing companies using the tale of the tape and looking to see who has the sword and shield of asymmetries on their side.
3. Watch a firm's important strategic choices that increase or decrease its chances of successfully managing the process of disruption.

The conclusion of each chapter listed a series of questions to help perform each step. Figure C-1 summarizes those questions.

FIGURE C-1

Questions in Analytical Process

Signals of Change	Competitive Battles	Strategic Choices
1. What jobs are customers in the industry trying to get done? Are customers not served, undershot, or overshot by current offerings? Along which dimension do firms compete for customers? 2. What improvements garnered premium prices in the past? 3. Do integrated or specialized business models currently prevail? Are interfaces specifiable, verifiable, and predictable? If so, where is modularity occurring? 4. Where are new business models emerging? Is there growth in fringe markets? 5. What role does the government or its regulatory bodies play in enhancing or inhibiting innovation?	1. What are industry players' business models? What are their motivations? What are their skills? 2. How do industry players compare to one another? How do they compare to the needs of the market? Where are there symmetries? Where are there asymmetries? 3. Do the asymmetries tilt in favor of the attacker or the incumbent? 4. Does the innovation naturally fit its target market? Is there evidence of cramming? 5. Are there signs that a company is ceding a low-end market and trying to move up? Is there an "up" to move to? For how long?	1. Is a company in a situation in which the right strategy needs to emerge? Is the firm giving itself the freedom to encourage emergent forces? Have managers wrestled with problems they are likely to face again? Have they shown the capacity to learn? 2. Are investor values aligned with company needs? If the investor is a corporation, has growth stalled? 3. Do value networks overlap? If they do, what are the degrees of overlap? Do they make it impossible to create a business model that has asymmetries? 4. Is this an appropriate situation for a spinout? Is the company giving the spinout the freedom to do what is necessary?

In chapters 5 through 8 and chapter 10 we showed how the analytical process provided insight into the future of the education, aviation, semiconductor, health care, and telecommunications industries. Each chapter's conclusion listed additional general lessons that emerged from each industry study. Figure C-2 summarizes those lessons.

Some of the most important lessons in *Seeing What's Next*, naturally, relate to disruptive innovations. Four critical lessons are:

1. Disruption is a process, not an event.

FIGURE C-2

Lessons from Each Chapter

Education

- Nonconsumption exists, even in contexts in which everyone seems to be consuming.
- Making it easier for nonconsumers to do things that matter to them is the ticket to growth.
- Companies can adopt the same innovation in very different ways.
- Companies can start at the fringes and then move to the core to avoid being snared in government regulation.
- All organizations—even nonprofit ones—have skills and motivation that drive what they can and cannot do.

Aviation

- Skills and motivation define companies' strengths and weaknesses, creating opportunities for new sorts of companies and determining companies' likely response to disruptive growth.
- Companies can create growth without incumbent response by leveraging asymmetric motivation.
- An industry's context can affect how companies react to disruptive threats.
- Companies with disruptive potential can choose co-optable business models by situating themselves in overlapping value networks.
- A company's choice of initial target market and value network can increase its disruptive potential.

Semiconductors

- The principle of conservation of integration helps describe how value chains need to reconfigure when the basis of competition shifts.
- Customers showing less interest in paying for performance increases and the entrance of nonintegrated firms that specialize at one stage of value added indicate overshooting.
- Overshooting enables companies to change the basis of competition, take advantage of rules-based design and manufacturing, and shift production closer to the customer.
- Disruptive business models are difficult for incumbents to respond to because they take unique advantage of blind spots and skill gaps.

Health Care

- Scientific progress leads to better categorization and the development of rules guiding prevention and treatment.
- Those rules open the door for less-skilled people to do what previously required deep expertise.
- Nonmarket forces affect the market for innovation by influencing industry players' motivation and ability.

Telecommunications

- The analytical approach laid out in chapters 1 through 3 can bring order to innovations in complicated industries.
- Theory can focus attention on the signs that signal an innovation's and an industry's evolution.
- Be wary of overestimating the chances that an innovation will overthrow incumbent leaders. Look at the co-optability of the innovation and the motivation of incumbents to fight the disruption rather than to flee it.
- Highly interdependent networked industries provide avenues for co-option.
- Developments in the fringe always bear watching.

2. Disruption is a relative phenomenon. What is disruptive to one company may be sustaining to another company.
3. Different or radical technology does not equal disruptive.
4. Disruptive innovations are not limited to high-tech markets. Disruption can occur in any product or service market and can even help explain competition among national economies.

We hope readers have improved their ability to identify when an innovator has developed a legitimate new-market or low-end disruptive innovation and to know what to watch for to make sure the innovator does not wander from the disruptive path.

Disruptive product or service innovations can create new markets by competing against nonconsumption or by bringing consumption to new contexts. These relatively simple, underperforming products allow a broader group of people to do things that only experts could do in the past. One common trap that companies with this type of innovation fall prey to is trying to force a disruptive innovation directly into a large, obvious market. Because of the product's initial limitations, the biggest, most lucrative customers will not initially be interested in purchasing the product. It requires a disciplined management team that has taken the right courses in the school of experience to avoid attempting to modify the innovation to go after the biggest market. That discipline is necessary for ultimate success. Attempts to cram a disruptive innovation into a large existing market almost never work. Customers tend to reject the innovation. The industry leaders are very motivated to stave off the attack and they have the resources and skills to do so.

Business model innovations can allow companies to offer goods or services more conveniently or at lower prices. These low-end disruptions also require disciplined management that limits initial deployment to the lowest market tier. Low-end disruptors find success when they stealthily take advantage of asymmetries of motivation by targeting customers whom existing competitors are happy to shed. Although head-to-head competition for the most valuable customers can be successful, it takes vast resources because incumbents will fight tooth and nail to hold on to their core markets.

Most innovations are not disruptive. Many of the most important and most profitable innovations are sustaining innovations that take a good product or service and make it better. Although incumbents typically lead sustaining innovations, entrants can succeed if they define success in the correct way. An entrant with a sustaining innovation has a low likelihood of success if it attempts to build a substantial business around the innovation. Unless the entrant has a lot of money to spend, this path is perilous. An entrant has a much higher likelihood of success if it defines success as working with an incumbent to produce a valuable augmentation that customers are willing to pay for. Often, the entrant ends up (profitably) selling its innovation to the incumbent.

Working with existing firms is especially important when a product or service is not good enough and improvement requires mastering unpredictable interdependencies. Existing firms that control important pieces of an industry's value chain are more useful as partners than as competitors. Entrants that try to go head-to-head against an incumbent in an industry that still demands integration are likely to disappoint customers. The entrants' product will not be as good as the incumbents' product. This is particularly true in the case of radical sustaining technologies. Existing firms seeking to move up-market can untangle the numerous new interdependencies and work out the inevitable kinks in the new technology. The scale and scope required to do this successfully are generally beyond an entrant company's reach.

The appropriate go-to-market strategy of both entrants and incumbents changes when overshooting occurs and interfaces shift in ways that encourage modularity. Specialist companies can enter the market at this stage and capture slices of value. Incumbents often presage the move to modularity by overshooting tiers of the market and trying to innovate faster to meet the needs of the market. However, incumbents often find that, whereas moves to modularity do help them innovate faster, value skates away from them to the suppliers that provide key performance-defining components and subsystems. Incumbents that recognize these shifts and act appropriately can skate to where attractive profits will lodge in the future.

Parting Ways

The journey does not end here, of course. The theory construction and utilization process is iterative. There are many ways for future researchers to build on and refine our work. Two particular areas of interest would be to look for anomalies (things the theories predict would happen that don't) and to develop deeper analytical tools (for instance, to develop a quantitative way to assess asymmetries).

Research along these lines will increase the usefulness of theory as a predictive tool. It is a never-ending quest. The future will always be unpredictable. The only thing we can truly predict with absolute certainty is that something we didn't predict will happen. Our goal is not—and cannot be—perfection. Our goal is to constantly increase our ability to interpret unpredictable developments. Think of the concepts presented in this book as a road map. When the unpredictable event inevitably happens, theory can help to quickly understand the event's implications on a country, industry, firm, or market segment.

As such, we will not judge the success of *Seeing What's Next* on the percentage of our predictions that come to pass. We will judge success by how useful readers find the concepts we have presented. We hope readers will develop the intuition to use our approach to gain insight into the future that matters to them by astutely identifying early signals and reacting accordingly. For some, it might mean understanding the implications of an innovation an ebullient coworker proudly described at the last departmental meeting. For others, it might mean seeing how to navigate through the choppy waters of a multibillion-dollar industry wrestling with wrenching change. For still others, it might mean assessing the growth prospects of an entire country.

Finally, five pieces of advice:

1. Don't be threatened by someone countering your insights by referring to "unassailable" data. Remember, truly unassailable data only exists about the past. Ask questions such as: "Are we sure this is the right data?" and "Isn't it possible that something here is different, so success in the past is not a useful predictor of success in the future?"
2. Don't pit theory and data as foes. Use theory to help guide data collection, looking for confirmation that circumstances are changing or indicators that suggest who has the upper hand in a competitive battle. As the old saying goes, "Trust. But verify."
3. Remember that everything is relative. The same innovation has very different implications for different companies. Evaluating an innovation's implications for a particular company requires understanding the company's strengths, weaknesses, mental

models, and operating protocols. Always seek to understand how the world looks through the company's eyes. Every company is good at some things and bad at others. Every company is motivated to tackle some opportunities and ignore others.

4. Remember the difference between announcements and actions. Just because a company says it plans to do something does not necessarily mean it *will*. Rarely do bold vision statements by senior managers that do not make sense to the rest of the organization turn into concrete actions. Companies often use press releases to create noise, but noise does not drive industry change.
5. Remember that choices matter—up to a point. Outcomes of competitive battles are rarely preordained from day one. Firms have great discretion to make decisions to enhance or lessen their chances of eventually emerging triumphant. But firms do not have unlimited degrees of freedom. Early decisions can greatly influence a firm's capabilities, which define its disabilities, which determine what strategic options will ultimately prove unpalatable.

Now, go use these theories. Search for anomalies and broadcast them so we can together improve our collective ability to see what's next.

APPENDIX

SUMMARY OF KEY CONCEPTS

After discussing the process of theory building, the appendix reviews the following concepts:

- Disruptive innovation theory
- Resources, processes, and values (RPV) theory
- Jobs-to-be-done theory
- Value chain evolution (VCE) theory (with corollary sustaining innovation classification scheme)
- Schools of experience theory
- Emergent strategy theory (with supporting discovery-driven planning tool)
- Motivation/ability framework

The description of each concept also references source material that explains the concept in more depth.

The Process of Theory Building

A theory is a contingent statement of what causes what and why. Business researchers typically follow a three-stage process (displayed in the figure following) to develop a theory. First, researchers carefully observe, describe, and measure phenomena. In simple terms, something is happening, and the researcher seeks to document it.

Process of Theory Building

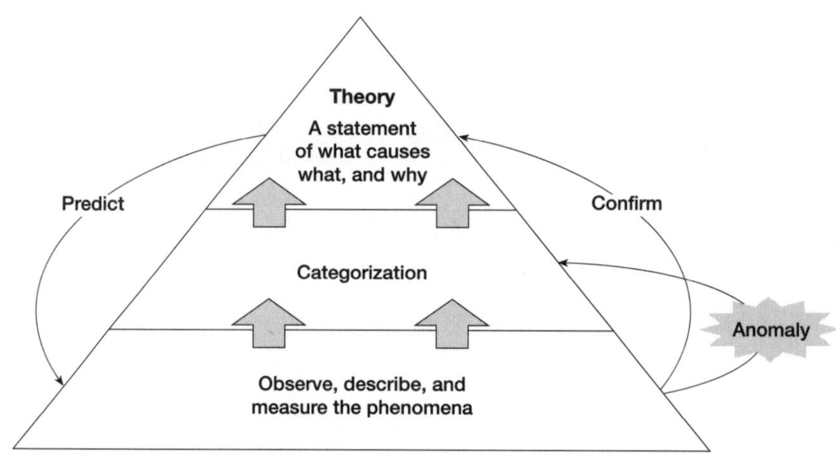

Then, to make life simple, people group the observations into distinct categories. They look for the most obvious similarities and differences between phenomena. Initially, researchers draw distinctions between categories based on the *attributes* of the phenomena. Attributes tend to be adjectives, describing things such as size and degree of difficulty. Then, researchers develop a theory that explains how a certain set of attributes leads to a certain result.

This is only the beginning of the theory-building process. After researchers develop an initial theory, they return to the bottom of the pyramid. They use the theory to predict the phenomena they will see in various situations. Typically, researchers encounter anomalies—things the theory predicts would not happen that do happen or vice versa. This is because theories based on attributes tend to show *correlation* but do not explain *causation*. The researcher cannot explain *why* different attributes lead to different results. Furthermore, other researchers who observe the same phenomena develop conflicting categorization schemes. This stage leads to significant confusion but is a necessary step on the path to robust theory. Although many researchers are taught to avoid anomalies, the discovery of anomalous phenomena is the pivotal element in the process of building improved theory.

Most breakthroughs in developing theories that have predictive power occur when researchers discover the fundamental mechanism that *causes* the phenomenon (getting beyond correlation) and then develop mutually exclusive and comprehensively exhaustive circumstance-based categories that reflect the different situations in which practitioners might find themselves. Circumstance-based categorization schemes allow researchers to understand why certain actions lead to certain results. The theory becomes a useful tool because it explains when actions will or will not lead to desired results and what changes make previously sensible actions unworkable. Getting the categories right is the key to developing robust, useful theory. If the same statement of cause and effect leads to the same outcome in two different situations, the distinction between categories is not meaningful.

It is important to draw a distinction between anomalies that a theory *cannot* account for and exceptions that a theory *can* account for. A theory with a good circumstance-based categorization scheme and a causal underpinning explains the actions a practitioner can take to change the circumstances or to account for the forces that act upon them.

For more information, see The Innovator's Solution, *chapter 1.*

Disruptive Innovation Theory

The concept of disruptive innovations came from Clayton Christensen's research into the rigid disk drive industry. That research identified more than 100 innovations and classified them as either *sustaining* to firms, in that they allowed the firms to provide better and more profitable products to their customers, or *disruptive*, because they offered initially poorer performance along the dimension that existing customers cared the most about. The findings indicated that incumbent companies almost always ultimately master sustaining innovations but often are unable to cope with disruptive innovations.

The following figure displays the disruptive innovation theory. A market is made up of many different groups of customers, whose needs are depicted in the figure as a series of dotted lines. These customers can be classified by how demanding they are—that is, what sorts of problems they are trying to solve and what types of products they require to solve those problems. For the sake of simplicity, the figure shows only a few customer groups. In reality, most industries contain numerous customer

The Disruptive Innovation Theory

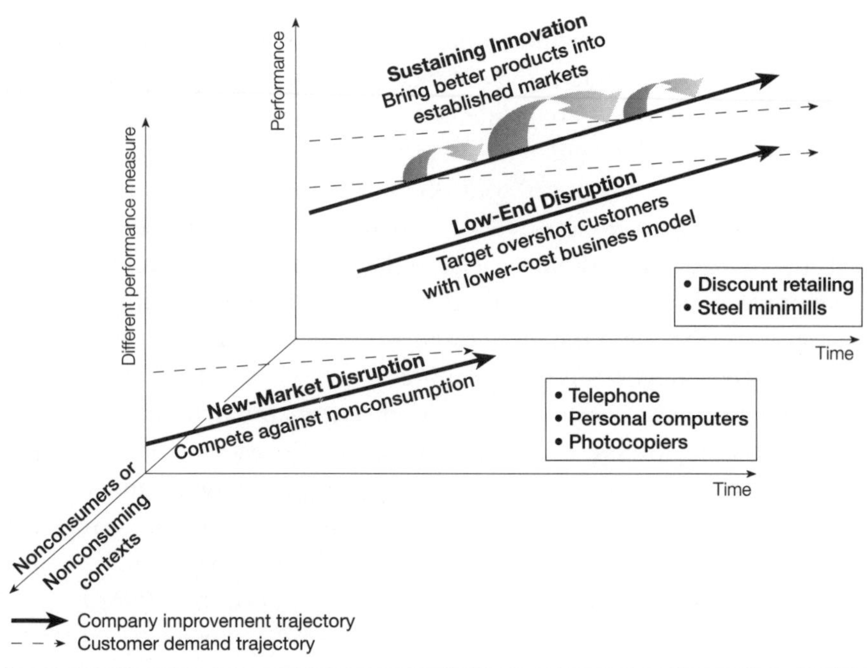

groups, comprising "tiers" in their markets, ranging from not very demanding to very demanding, with the great majority of customers somewhere in the middle. This majority of customers is termed the *core* of the market, or the *mainstream* customer.

Companies have very strong incentives to innovate and move upmarket because more demanding customers tend to pay premium prices for products or services that solve the toughest problems. Well-run companies that listen to their best customers—as they have been repeatedly taught—develop sustaining innovations that improve their products along the dimensions that customers have historically valued. These sustaining innovations essentially take a good product and make it better and are crucial to a company's growth and prosperity.

However, firms almost always improve their products faster than customer needs can change to use the new innovations. Therefore, incumbent firms tend to create new products and services at a pace (illustrated by the dark arrows in the figure) that outstrips the ability of customers in various levels or tiers of the market to use the improvements.

Disruptive products or services initially are inferior to existing offerings in the attribute that matters most in the mainstream. However, they are typically more affordable and simpler to use than products in the incumbent's product portfolio. All disruptions are predicated on creating growth opportunities away from the core of the incumbent's market. Companies seeking to build disruptive businesses can capitalize on two distinct types of opportunities. They can seek to *compete against nonconsumption* and establish a completely new market (new-market disruption), or they can *attack from the low end* by deploying a business model that profitably serves the less demanding customers whom market leaders are happy to shed as they themselves move up-market (low-end disruption).

Initially, the disruptive innovation is not good enough to meet the performance requirements of the core market. However, the disruptive innovators aggressively move up-market on their own sustaining improvement trajectories as they pursue more attractive profit margins. Ultimately, when the disruptive innovation is good enough to meet the needs of larger swaths of the incumbent's customers, the incumbent is forced further up-market or out of the market entirely.

Disruptive innovations typically pose difficulties for an industry's incumbent leaders, who are faced with the choice of investing to improve their products along the sustaining trajectory or investing in disruptive innovations. The choice is difficult because sustaining innovations appear to be much more attractive than disruptive ones, even though disruptive innovations often ultimately drive more long-term growth.

For more information, see The Innovator's Dilemma *and* The Innovator's Solution, *chapters 2 and 4.*

Resources, Processes, and Values Theory

The resources, processes, and values (RPV) theory explains why existing companies tend to have such difficulty grappling with disruptive innovations. The RPV theory holds that resources (what a firm has), processes (how a firm does its work), and values (what a firm wants to do)

Resources, Processes, and Values

Resources	Processes	Values
Things or assets that organizations can buy or sell, build or destroy. Examples: • People • Technology • Products • Equipment • Information • Cash • Brand • Distribution channels	Established ways companies turn resources into products or services. Examples: • Hiring and training • Product development • Manufacturing • Planning and budgeting • Market research • Resource allocation	The criteria by which prioritization decisions are made. Examples: • Cost structure • Income statement • Customer demands • Size of opportunity • Ethics

collectively define an organization's strengths as well as its weaknesses and blind spots.

Resources (depicted in the left column of the figure above) are things or assets that organizations can buy or sell, build or destroy. Processes (depicted in the middle column of the figure above) are the established patterns of work by which companies transform inputs of resources into outputs—products, services, or other resources—of greater worth. Values (depicted in the right column of the figure above) determine the criteria by which organizations allocate their resources.

The RPV theory argues that organizations successfully tackle opportunities when they have the resources to succeed, when their processes facilitate what needs to get done, and when their values allow them to give adequate priority to that particular opportunity in the face of all other demands that compete for the company's resources. Incumbent firms master sustaining innovations because their values prioritize them and their processes and resources are designed to tackle precisely those types of innovations. Incumbent firms fail in the face of disruptive innovations because their values *will not* prioritize disruptive innovations and the firm's existing processes do not help them do what needs to get done.

For more information, see The Innovator's Solution, *chapter 7, and* The Innovator's Dilemma, *2d ed., chapter 8.*

Jobs-to-Be-Done Theory

When consumers buy a product, they are really hiring the product to get a job done for themselves. The jobs we are seeking to get done in life are remarkably stable. Companies are successful when they make it easier for their customers to get done something they historically cared about.

Market segmentation is a theory. Companies think that meeting the needs of the members of a segment will lead to a successful new product. Unfortunately, companies tend to base their segments along attributes that are easy to identify and measure, such as product characteristics, demographics, or psychographic profiles. However, these attribute-based segments tend to be inadequate. Drawing arbitrary lines around groups defined by easy-to-gather data obfuscates customers' real needs.

Faulty market segmentation schemes help to explain the stunningly high rate of failure of new product development. We just don't live our lives in product categories or in demographics. When companies segment markets this way, they often fail to connect with their customers.

The jobs-to-be-done theory holds that products are successful when they connect with a circumstance—with a job that customers find themselves needing to get done. Products that successfully match the job or the circumstance we find ourselves in end up being the real "killer applications." They make it easier for consumers to do something they were already trying to accomplish.

By identifying what jobs people really care about and developing products that make it easier to achieve these jobs, companies can identify new markets that they were previously unaware of and that could not be uncovered via traditional market research. Understanding those jobs that are not adequately satisfied by current products provides deep insight into what are and what will be the truly innovative products that delight existing customers and attract new customers from the sidelines of nonconsumption.

For more information, see The Innovator's Solution, *chapter 3.*

Value Chain Evolution (VCE) Theory

Industries tend to evolve from states of *interdependence*, where leading firms need to be vertically integrated, to *modularity*, in which specialist firms that are responsible for critical pieces of the value chain and

The Value Chain Evolution Theory

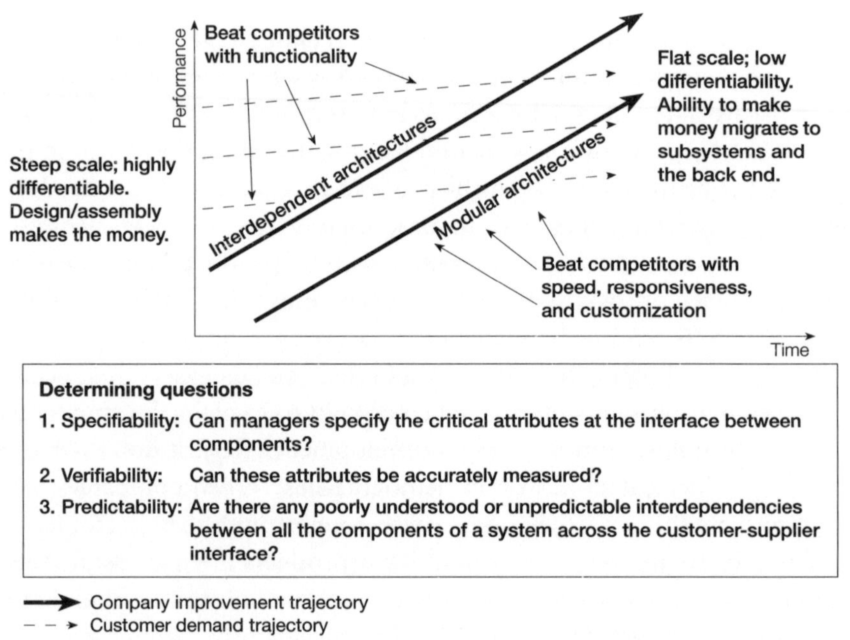

produce key product components can earn a disproportionate share of value in an industry.

Before a product or service is good enough to meet mainstream customer needs (see the left side of the figure above), integrated firms that control the entire production and delivery process are best suited to coordinate the complexities developers will confront when trying to improve the product. Companies that solve these problems are rewarded with a disproportionate share of industry profit. To improve products, companies often must use new and unproven technologies or put existing technologies together in new ways, creating new patterns of interaction and new problems. Pushing the frontier of what is possible requires proprietary, interdependent architectures, and in this circumstance, integration gives firms a full platform to run engineering experiments to wring out continued improvements.

Stitching together a system with a series of partner companies is next to impossible when there are complex interactions across the boundaries of what the firms provide. Management is the only force capable of coordinating these interdependencies. Thus integrated firms tend to earn the lion's share of industry profit in circumstances in which a product's functionality and reliability are not yet good enough.

As companies overshoot their customers' needs (see the right side of the previous figure), companies no longer need the benefits that integration brings. Instead, they increasingly compete based on speed, flexibility, or convenience. In an effort to develop products or services more quickly, companies tend to standardize interfaces between various parts of the product or service. These standards eventually morph into industry-wide standards and allow product architecture to become modular. Modular products or services allow companies to get to market more quickly because they can replace individual components without redesigning an entire product. Modularity also enables the creation of specialist firms capable of developing products that fit these interfaces. This change allows previously integrated firms to outsource pieces of their product to vendors that meet their specifications.

Three questions determine whether an interface truly can become modular:

1. Managers must be able to *specify* what aspects of the interface are important and which are not. If managers cannot articulate and document their requirements to a supplier so the supplier will know how to easily interconnect, it is not possible to "peel off" a piece of the product.
2. Managers must be able to *verify* that the exchanges occurring across this interface are correct and are what they need. This means that whatever aspects have been specified must also be measurable.
3. The interactions across the interface must be well understood and *predictable*. If there are unpredictable interactions between components, efforts to move toward modularity have potentially disastrous consequences.

If the transactions that occur across an interface between components are specifiable, verifiable, and predictable, specialist firms can understand and successfully manage the related interactions.

For more information, see The Innovator's Solution, *chapters 5 and 6.*

Corollary: Sustaining Innovation Classification Scheme

Sustaining innovations play two important roles in industry change: They define the path of incumbent improvement and they provide the fuel for disruptive companies to march up their own improvement trajectories. The sustaining innovation classification scheme (see the figure below) is a corollary to the disruptive innovation and VCE theories. It identifies three unique circumstances that lead to three different types of sustaining innovations, each of which affects an industry in a unique manner.

Sustaining Innovation Classification Scheme

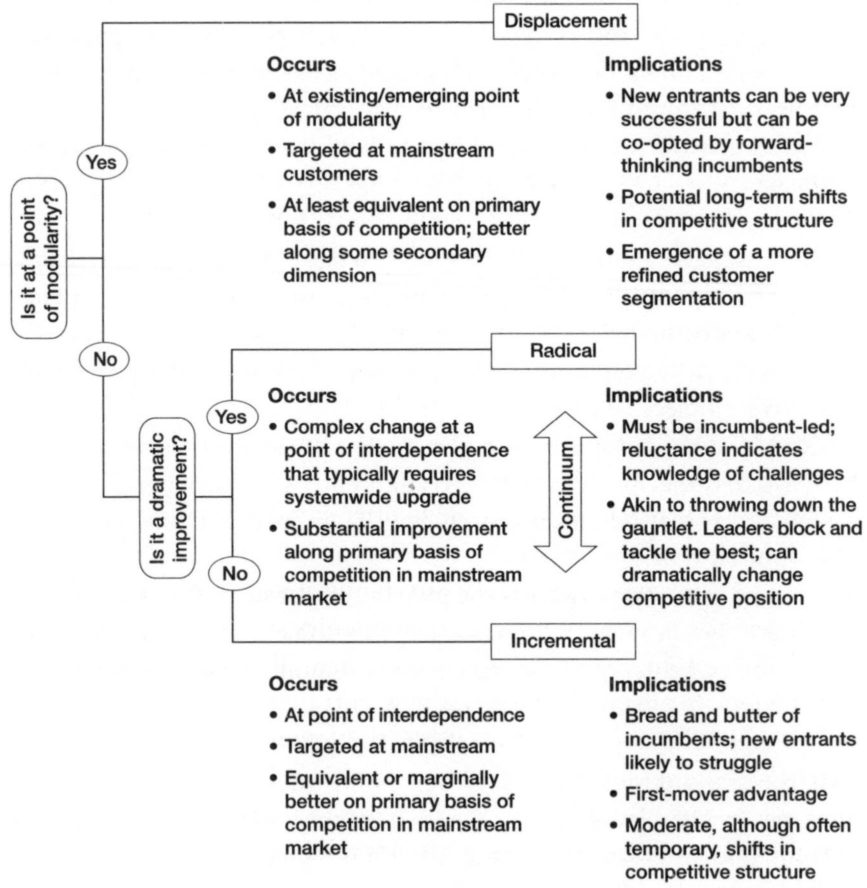

The first question to ask when seeking to classify a sustaining innovation is: Does the innovation occur at a point of modularity? If the answer is yes, the innovation is a displacement. Displacements are innovations that target a specific piece of an industry's value chain. Specialist entrants can introduce displacements, but forward-thinking incumbents can co-opt displacements.

If an innovation does not take place at a point of modularity, it is one of the two other forms of sustaining innovations—radical or incremental—that both happen at points of interdependence. Radical sustaining innovations are at the complex end of the continuum. These "great leaps forward" tend to be very complicated and expensive. Only integrated incumbents who control large swaths of an industry's value chain can introduce radical sustaining innovations. Integrated companies can master the myriad interdependencies involved in wrestling with compatibility, interoperability, and legacy issues. Specialist companies just don't control enough pieces of the puzzle to make their radical solutions good enough for the market's needs.

Radical sustaining innovations give incumbent firms an opportunity to dramatically change their relative competitive positions in a marketplace. Launching a radical sustaining innovation is akin to throwing down the gauntlet, saying, If you want to play in the game, you have to pay the price and upgrade. Firms can improve their competitive positions by doing the best job of "blocking and tackling" and managing the complicated implementation process. Being a fast follower can be a very successful strategy. The first mover will have to master many novel and unpredictable technologies. The followers can potentially avoid these expenses.

Incremental sustaining innovations offer smaller improvements than radical sustaining innovations. They tend to influence an industry less dramatically. Because incremental sustaining innovations still occur at interdependent interfaces, integrated companies still have an advantage. In fact, developing innovations that incrementally improve products is what incumbent firms do best. If a new entrant attempts to introduce an incremental sustaining innovation, the incumbent has both a will and a way to respond. Firms that first introduce incremental sustaining innovations can gain a temporary increase in market share. However, incremental sustaining innovations do not tend to cause seismic market-share shifts.

Schools of Experience Theory

The schools of experience theory holds that managers are more likely to be successful when they encounter problems that they wrestled with in the past. This is because managers learned strategies and methods to grapple with problems when they encountered those problems in past assignments. These problems constitute courses that managers have taken in the school of experience.

Even though hiring is perhaps the most important task facing operating managers, their success rate is astonishingly low. Manufacturers around the world have adopted Six Sigma processes that attempt to remove 99.999 percent of errors from all production processes, but managers candidly admit that roughly one out of four people they hire or promote turns out to be a mistake. Manufacturers would not tolerate a 25 percent failure rate on their production lines, so why do managers tolerate it in their hiring decisions?

Many managers respond by saying that hiring is an art, not a science. Therefore, we should expect and tolerate high failure rates. Ironically, operations management experts used the same argument years ago to explain why the production process had intrinsic randomness. But perceived randomness disappeared once researchers understood the important variables behind the randomness.

One explanation for hiring's high failure rate is the overwhelming emphasis on finding the *right stuff*. Right-stuff thinkers search for people with the right attributes. Evidence of the right stuff might appear in the educational institutions a candidate attended, her success in managing complex problems in another company, or in other evidence of a record of accomplishment under difficult circumstances.

However, most research indicates that successful executives are *made* more than they are *born*. Most of what managers have learned results from wrestling with problems. Overcoming these challenges or problems leads to the development of competencies that people can use in similar situations in the future. While right-stuff thinking castigates failure, failing can actually be beneficial as long as one can learn to identify the root cause of the failure so as to prevent it in the future.

When hiring, organizations need to look past adjectives that describe people. They need to focus on the past tense verbs on a candidate's résumé that indicate the challenges that candidate faced in his or her past.

Those verbs ought to match up with the problems the organization knows it will confront.

For more information, see The Innovator's Solution, *chapter 7, and McCall,* High Flyers.[1]

Emergent Strategy Theory

The emergent strategy theory holds that in highly uncertain circumstances, companies need to develop ways to adapt to marketplace signals.

Companies have two different ways to set strategy. They can follow a *deliberate* strategy, where they set a goal, define a set of steps to reach that goal, and then methodically act on each step. This process is very conscious and typically quite analytical. It involves assessment of market structure, competitive analysis, and detailed market research to determine customer needs.

Companies that take the other approach—following an *emergent* strategy—try to retain flexibility and gather feedback from the marketplace on what works and what doesn't. They try to change their strategies on the fly to adapt to new information that emerges from the marketplace.

Emergent strategies work in highly uncertain situations. In these situations, operating managers tend to encounter problems that business planners didn't anticipate. Actions lead to unanticipated results. In such circumstances, following a rigorous deliberate strategy can lead companies to ignore market signals and not adapt their strategies. In other words, they can continue to stick to a strategy that clearly isn't working. Emergent strategies encourage managers to respond to problems in the most appropriate way, even if it results in deviating significantly from the deliberate course.

In reality, strategy making is a complicated process that blends both deliberate and emergent strategy. Even firms that try to follow a purely deliberate strategy-making process surprisingly find that their actual strategy diverges quite significantly from their intended strategy because of the variety of emergent forces. Firms often struggle when they try to shut off these emergent forces before it is clear that they know the right approach.

For more information, see The Innovator's Solution, *chapter 8, and Burgelman,* Strategy Is Destiny.[2]

Supporting Tool: Discovery-Driven Planning

Discovery-driven planning is an important tool to help companies implement the emergent strategy theory. The rigidity of the typical planning process inevitably creates conflict between the direction a firm is moving toward and its employees' everyday decisions. A company needs to have a planning process that can incorporate the output of an emergent strategy-making process into an effective and appropriate allocation of resources. Companies can use discovery-driven planning as a tool to accommodate emergent forces.

The typical planning process involves creating a set of assumptions, developing projections based on those assumptions, building a plan based on those projections, and then acting on the plan. This planning process works extremely well in established markets where companies can draw on a great wealth of data and accumulated experience.

In uncertain situations, companies do not need to fly by the seat of their pants. They can follow a rigorous process by using discovery-driven planning. Discovery-driven planning (see the figure below) flips the planning process on its head. It starts with projections—what must an opportunity look like in order for it to be an interesting investment possibility? The next stage is to map out what assumptions would need to prove true for the projections to happen. With these codified assumptions, a company can consciously experiment to test the assumptions

Discovery-Driven Planning

	Platform-Based	Discovery-Driven
Step 1	Make assumptions	Make *projections*
Step 2	Build projections based on assumptions	Determine assumptions that must prove true in order for projections to happen
Step 3	Make decisions to invest based on projections	Implement a plan to *learn*—to test whether the critical assumptions are reasonable
Step 4	Implement the strategy	Invest to implement the strategy

rather than assume the unknowable. As assumptions get tested, some will bear out and some will not. Discovery-driven planning allows a company to test uncertainties and develop contingent plans before reaction is too late.

For more information, see The Innovator's Solution, *chapter 8, and McGrath and MacMillan,* "Discovery-Driven Planning."[3]

The Motivation/Ability Framework

The motivation/ability framework (see the figure below) helps assess how nonmarket forces affect innovation. The framework shows that there are two necessary inputs to innovation. The first input is motivation, or market incentives. The second input is ability, or the capability to obtain resources, craft them into products and services, and offer

The Motivation/Ability Framework

Generally determined by
- Market size/growth
- Competitive dynamics/ industry attractiveness
- Economics of opportunity/ business model attractiveness
- Competitive forces

Government levers
- Tax treatment (credit, subsidies, etc.)
- Antitrust policy
- Competitive policy
- Rate regulation
- Regulatory asymmetry
- Network element pricing

Motivation (Low – High)

	Low Ability	High Ability
High Motivation	**Looking for a Target** Firms constrained in accessing resources or reaching potential customers	**The Hotbed** Teeming with innovation
Low Motivation	**The Dilemma** No readily available avenues to create profitable businesses	**Looking for the Money** Firms struggle to find ways to monetize an opportunity

Ability (Low – High)

Generally determined by
- Resource availability
- Standards
- Market access
- Industry development

Government levers
- Resource-related regulation
- Unbundling
- Standards

those products and services to customers. Markets that have high levels of both ability and motivation tend to have high levels of innovation.

Nonmarket factors, such as industry standards, unions, cultural norms, the state of technological development, a country's intellectual property infrastructure, and, most important, government regulation affect the motivation and ability to innovate.

Simply stated, the motivation/ability framework suggests innovation flourishes when companies have both the motivation and ability to innovate. Market contexts that are deficient in motivation or restrict ability are stifling to innovation. Companies that bring nascent innovations to unfavorable market environments end up either searching for environments that are more favorable or abandoning their efforts.

Government's power to affect innovation lies in its policymaking and regulatory authority. The government or other nonmarket players can affect either the motivation or ability of industry participants. In doing so, they can change an industry's context, making it more or less conducive to innovation.

Using the framework can point to barriers slowing the pace of innovation. Nonmarket players can improve the market for innovation by taking actions to remove these barriers. Nonmarket players can damage the market for innovation by erecting new barriers. Efforts by nonmarket players struggle when they address the wrong barriers.

For more information, see Seeing What's Next, *chapter 4.*

Notes

1. Morgan W. McCall Jr., *High Flyers: Developing the Next Generation of Leaders* (Boston: Harvard Business School Press, 1998).

2. Robert A. Burgelman, *Strategy Is Destiny: How Strategy-Making Shapes a Company's Future* (New York: Free Press, 2002).

3. Rita Gunther McGrath and Ian C. MacMillan, "Discovery-Driven Planning," *Harvard Business Review*, July–August 1995, 44–56.

GLOSSARY

Ability (motivation/ability framework; signal of change): One of the two drivers of innovation that nonmarket forces can influence. Defined as the capability to obtain resources, craft them into a business model, and offer them to customers. Generally determined by the scarcity of resources, standards, industry development, and market access. Key nonmarket levers include resource-related regulation, unbundling, standards, and approval mechanisms. When nonmarket players take action that affects ability, it is a signal of change. *See* motivation

Architecture (VCE theory): Determines a product's constituent components and subsystems and defines how they must fit and work together to achieve the targeted performance.

Asymmetric motivation (competitive battles): When a firm does something that another firm does not want to do. Asymmetric motivations shields entrants from competitive responses in their early stages. Entrants using a disruptive strategy will enter small markets with low margin dollars per unit. Incumbents are motivated to ignore such markets. Also powers incumbent flight to up-market opportunities. Markets that look great to companies at an industry's low end look terrible to companies at an industry's high end. *See* asymmetric skills; sword and the shield

Asymmetric skills (competitive battles): When a firm does something that another firm is incapable of doing. Asymmetric skills explain why disruptors ultimately triumph over incumbents. Enduring asymmetric skills are rarely found in resources, because most resources can be hired or acquired. The asymmetries of skills that are hardest to replicate are rooted in the processes and business models that disruptive firms have honed as they have delivered on the unique needs of their home markets. *See* asymmetric motivation; sword and the shield

Attribute (process of theory building): A characteristic of phenomena along which researchers create categories. Attribute-based categorization schemes tend to lead to theories that are statements of correlation, not causation. *See* circumstance; theory

Basis of competition (VCE theory): The dimension of innovation along which improved products and services garner premium prices from customers. Typically goes from functionality to reliability to convenience to price. *See* good enough; not good enough

Business model: The way a company captures value from its innovations. This includes the structure of its costs, how it prices its product or service, whom it attempts to sell that product or service to, how it sells it (one-time sale, licensing agreement, and so on), what value proposition it purports to offer, how it delivers its product or service, how it offers postsales support, and so on.

Circumstance (process of theory building): A situation in which practitioners can find themselves. Researchers who create circumstance-based categorization schemes can develop theories that are contingent statements of causality. *See* attribute; theory

Conservation of integration (VCE theory): A corollary to the value chain evolution theory. It holds that when an interdependent architecture is necessary to optimize performance at a stage of value added that is not good enough, the architecture of the product or service at the adjacent stage of value added must be modular and conformable, in order to optimize the performance of what is not good enough. This means that when the architecture at one stage of value added goes from interdependence to modularity, the architecture of the adjacent stage of value added is likely to go from modularity to interdependence.

Co-option (competitive battles): Incumbent companies incorporating potentially disruptive innovations into their existing businesses. Co-option becomes much more likely when asymmetries do not exist.

Cramming (competitive battles): Occurs when companies try to stretch an underperforming disruptive innovation to meet the needs of demanding customers in mainstream markets. Cramming is expensive, leads to disappointed customers, and generates little new growth.

Deliberate strategy (strategic choices): A top-down strategy that is formulated in a project and implemented by senior management. *See* emergent strategy; preparation regimen

Displacing sustaining innovation (signal of change): An innovation introduced by a specialist firm at a point of modularity that accompanies the disintegration of an industry. Displacing innovations often, but not always, enable disruptive business models.

Disruptive black belt (strategic choices): An incumbent firm learning to harness the forces of disruption either by appropriately creating spinout organizations or by developing internal skills to create successive disruptive innovations.

Disruptive innovation (disruptive innovation theory): An innovation that cannot be used by customers in mainstream markets. It defines a new performance trajectory by introducing new dimensions of performance compared to existing innovations. Disruptive innovations either create new markets by bringing new features to nonconsumers or offer more convenience or lower prices to customers at the low end of an existing market. *See* low-end disruptive innovation; new-market disruptive innovation; sustaining innovation

Emergent strategy (strategic choices): A bottom-up strategy that evolves and adapts based on signals that emerge from the marketplace. *See* deliberate strategy; preparation regimen

Entrant: A newly established firm within a market segment. A relative term; a firm with fifty years of history still can be an entrant if it enters a new market segment. *See* incumbent

Good enough (VCE theory, disruptive innovation theory): Performance that satisfies the basic needs of a market segment along a particular dimension, typically the basis of competition. *See* basis of competition; overshot customers

Incremental sustaining innovation (signal of change): An innovation that provides marginally improved performance along an established performance trajectory. *See* radical sustaining innovation

Incumbent: An established firm within a market segment. Established firms that go to new markets are considered entrants in that market segment. *See* entrant

Innovation: Anything that creates new resources, processes, or values or improves a company's existing resources, processes, or values. Obvious innovations include new or improved products, processes, and services. New delivery mechanisms, customer service strategies, and business models are all other forms of innovation.

Integration (VCE theory): When a company controls adjacent stages of value added in a product or service architecture. Full integration involves end-to-end production and delivery of a product. Firms can be partially integrated. *Operative integration* occurs when a firm not only owns assets in adjacent stages of value added, but also has interdependent processes that span across those stages.

Interdependent interface (VCE theory): An interface between parts, in which one part cannot be created independently of another part because the way one is designed and made depends on the way the other is designed and made. When there is an interface across which there are unpredictable interdependencies, the same organization must simultaneously develop both components if it hopes to develop either component. *See* modular interface

Interface (VCE theory): The place where any two components come together. Interfaces exist within a product as well as between stages in the value-added chain.

Job to be done: A circumstance in which an individual is trying to solve a problem or accomplish a task. Companies find success when they make it easier or more convenient for customers to do jobs that they had already been trying to do. A product usually fails when managers expect their customers to prioritize something that they historically had not been trying to do.

Low-end disruptive innovation (disruptive innovation theory, signal of change): An innovation that offers overshot customers good-enough performance along traditional metrics at lower prices. Supported by a business model that makes attractive returns at discount prices. *See* disruptive innovation; new-market disruptive innovation; overshot customers

Modular interface (VCE theory): A clean interface with no unpredictable interdependencies across components or stages of the value chain. Modular components can be developed by independent work groups or companies and still fit and work together in well understood and highly defined ways. *See* interdependent interface

Motivation (motivation/ability framework, signal of change): One of the two drivers of innovation that nonmarket forces, such as regulatory agencies, can influence. Defined as a pot of gold or market incentives. Generally determined by market size, competitive dynamics, economics of opportunity, and competitive forces. Nonmarket levers include rate regulation, regulatory asymmetry, tax treatment, antitrust policy, and competitive policy. When nonmarket players take actions that affect motivation, it is a signal of change. *See* ability

New-market disruptive innovation (disruptive innovation theory, signal of change): An innovation that makes it easier for nonconsumers to do certain jobs themselves or to do jobs in more convenient, decentralized locations. Typically has limitations along metrics valued by existing customers but provides benefits along new attributes such as convenience, customization, or low prices. *See* disruptive innovation; low-end disruptive innovation; nonconsumption

Nonconsumption (disruptive innovation theory, signal of change): An absence of consumption. Typically refers to either people (nonconsumers) or contexts (nonconsuming contexts). Nonconsumption occurs when attributes of existing products or services limit consumption to those with wealth or specialized training. New-market disruptions begin by competing against nonconsumption. Relatively simple innovations can trump nonconsumption. When companies reach nonconsumers or nonconsuming contexts, it is a signal of change. *See* new-market disruptive innovation

Not good enough (VCE theory, disruptive innovation theory): Performance that does not satisfy the basic needs of a market segment along a particular dimension, typically the basis of competition. *See* basis of competition; undershot customers

Overshot customers (VCE theory, disruptive innovation theory, signal of change): A particular customer segment for which existing products or services are more than good enough. Companies can reach these customers with low-end disruptions. Modular interfaces tend to emerge once overshot customers emerge, allowing companies to follow defined rules to develop good-enough products. When companies develop new ways to reach overshot customers, it is a signal of change. *See* good enough; low-end disruptive innovation; undershot customers

Preparation regimen (strategic choices): The decisions an entrant makes as it puts into place its initial set of resources, processes, and values. Specific critical decisions include sourcing financing, selecting a strategy-making process, and hiring. Getting initial conditions right is critical, because those conditions determine how attractive the subsequent decisions that need to be made correctly will appear. Choosing the wrong preparation regimen can lead to the creation of co-optable business models, among other things. *See* deliberate strategy; emergent strategy; schools of experience

Processes (RPV theory): The patterns of interaction, coordination, communication, and decision making that employees use to transform inputs of resources into products, services, and other resources of greater worth. Processes develop when a group repeatedly works together to perform recurring tasks or solve hard problems. Because it is meant to achieve the same acceptable results time after time, a process is inherently inflexible. The same process that constitutes a strength in addressing one type of task, therefore, typically constitutes a weakness if it is used to address a different type of task. Asymmetric skills are akin to asymmetric processes. Key processes include hiring and training, product development, manufacturing, planning and budgeting, market research, and resource allocation. *See* resource-allocation process; tale of the tape

Radical sustaining innovation (signal of change): An innovation that provides dramatically improved performance along an established performance trajectory. Typically requires managing messy interdependencies. *See* incremental sustaining innovation

Resource-allocation process (RPV theory): The process by which firms allocate resources to different opportunities. The resource-allocation process is a diffused, difficult to control process because it is comprised of prioritization decisions that are made autonomously by many people in an organization. Senior management directly controls *some* funding decisions. Other resources, however, are allocated to certain priorities and withheld from others, in decisions made by engineering managers, salespeople, human resources personnel, and so on. The resource-allocation process plays a key role in disruption because it leads incumbent firms to underinvest in disruptive opportunities. *See* processes

Resources (RPV theory): Tangible and intangible things a company has access to. Tangible resources include employees, technology, products, items in the balance sheet, equipment, customer relationships, and distribution network. Intangible resources include human capital, brands, and accumulated knowledge. *See* tale of the tape

Schools of experience (strategic choices): The set of challenges or problems that a manger has wrestled with in earlier assignments. Each one of these problems can be thought of as a "course" that the manager has taken in the school of experience. By taking these courses, managers develop the skills to successfully address the same challenges again in the future. *See* preparation regimen

Sustaining innovation (disruptive innovation theory): An innovation that moves a company along an established performance trajectory by introducing improved performance compared to existing products and services. Sustaining innovations can be radical or incremental. Displacing innovations also, on occasion, can have a sustaining impact. *See* disruptive innovation

Sword and the shield (competitive battles): Language to use to help visualize how to assess whether a firm is hiding behind asymmetric motivation (the shield) and attacking with asymmetric skills (the sword). *See* asymmetric motivation; asymmetric skills

Tale of the tape (competitive battles): Technique to evaluate a firm's resources (what it has), processes (what it can do), and values (what it wants to do). *See* processes; resources; values

Theory: A situation-specific statement of cause and effect. The best theories are predicated on robust circumstance-based categorization schemes that provide a guide to the situations managers encounter. This underpinning helps managers understand how different actions may be needed in different circumstances in order to achieve the needed results. *See* attribute; circumstance

Undershot customers (VCE theory, disruptive innovation theory, signal of change): A particular customer segment for which existing products or services are not yet good enough. *See* not good enough; overshot customers

Value network (strategic choices): The collection of upstream suppliers, downstream channels to market, and ancillary providers that support a common business model within an industry. When would-be disruptors enter into existing value networks, they must adapt their business models to conform to the value network and therefore fail at disruption because they become co-opted. Entrants have a greater chance of success if they seek out freestanding value networks. In the diagram of disruption, which shows the intersecting trajectories of improvement that innovating companies provide versus the improvement that customers can utilize, a new value network is depicted in the third dimension, as a new plane in which the unique value network coalesces.

Values (RPV theory): The criteria that employees use in a company's resource-allocation process to determine which opportunities get prioritized. Key elements of values include the way a company makes money, its cost structure, size, growth expectations, ethics, and mission. *See* tale of the tape

INDEX

ABOUT THE AUTHORS

Clayton M. Christensen is the Robert and Jane Cizik Professor of Business Administration at Harvard Business School. He holds a joint appointment in the Technology and Operations Management and General Management faculty groups. His research and teaching interests center on the management of technological innovation and finding new markets for new technologies. Prior to joining the HBS faculty, Christensen served as Chairman and President of CPS Corporation, a materials science firm he cofounded with several MIT professors. He was also a White House Fellow in the administration of President Ronald Reagan, and was a member of the staff of the Boston Consulting Group.

Christensen is the author of numerous publications, including the landmark books *The Innovator's Dilemma* and *The Innovator's Solution.* He serves as a consultant to the management teams of many of the world's leading corporations, and serves as a member of The Church of Jesus Christ of Latter-day Saints in as many ways as he can.

Christensen holds a B.A. in economics from Brigham Young University, an M.Phil. in economics from Oxford University, where he studied as a Rhodes Scholar, and an M.B.A. and D.B.A. from Harvard Business School. He and his wife, Christine, are the parents of five wonderful children: Matthew and his wife Elizabeth, Ann, Michael, Spencer, and Katie.

Scott D. Anthony is a Partner at Innosight, a management consulting and education company based in Watertown, Massachusetts. Founded by Clayton M. Christensen in 2000, Innosight seeks to help companies take advantage of disruptive innovation to create new growth. At Innosight, Anthony has worked with clients ranging from national governments to leading medical device, health care, chemicals, telecommunications, software, publishing, and consumer products companies. He has played an integral part in developing Innosight's methodology to implement the ideas in *The Innovator's Solution*. He also serves as the Editor of *Strategy & Innovation*, a journal that publishes cutting-edge thinking on these topics, coproduced by Innosight and Harvard Business School Publishing.

Prior to joining Innosight, Anthony worked with Christensen at Harvard Business School, managing a group that worked to further Christensen's research on innovation. He has authored a number of articles, cases, and working papers. Anthony has also worked as a Consultant for McKinsey & Co., a Strategic Planner for Aspen Technology, and a Product Manager for WorldSpace Corporation.

Anthony received a B.A. in economics, *summa cum laude*, from Dartmouth College and an M.B.A. with high distinction from Harvard Business School, where he was a Baker Scholar. He lives in Boston in the shadows of Fenway Park with his wife, Joanne.

Erik A. Roth is a Consultant in the Boston office of McKinsey & Co. While at McKinsey, Roth has focused his client work on the consumer packaged goods, retail, and telecommunications industries.

Prior to joining McKinsey, Roth worked with Clayton M. Christensen at Harvard Business School, where he helped further Christensen's research on innovation. He also authored a number of case studies and working papers. Previously, Roth spent several years as Marketing and New Product Development Manager for OnStar, General Motors's in-vehicle wireless telecommunications service. He was instrumental in the launch of new technology platforms and services while working to grow OnStar's subscriber base to 1 million. Roth also worked for a number of years as a Consultant for the Marketing Corporation of America, where he advised *Fortune* 500 clients on marketing and strategic growth issues focusing on the consumer packaged goods and high-tech industries.

Roth received a B.A. with high honors in international politics and economics, *cum laude*, from Middlebury College, and an M.B.A. from Harvard Business School. He lives in Cambridge, Massachusetts, with his wife, Kate, and his daughter, Annabel.